KU-318-411

Digital Filter Design

TOPICS IN DIGITAL SIGNAL PROCESSING

C. S. BURRUS and T. W. PARKS: *DFT/FFT AND CONVOLUTION ALGORITHMS*
Rice University

T. W. PARKS and C. S. BURRUS: *DIGITAL FILTER DESIGN*
Cornell University and Rice University

J. TREICHLER, R. JOHNSON, JR. and M. LARIMORE: *THEORY AND DESIGN OF ADAPTIVE FILTERS*
Cornell University and Applied Signal Technology, Inc.

W. KOHN: *DIGITAL CONTROL (in preparation)*
Rice University

Digital Filter Design

T. W. Parks
School of Electrical Engineering
Cornell University
Ithaca, New York 14853

C. S. Burrus
Department of Electrical and Computer Engineering
Rice University
Houston, Texas 77251

A WILEY-INTERSCIENCE PUBLICATION
JOHN WILEY & SONS, Inc.
New York   Chichester   Brisbane   Toronto   Singapore

Copyright © 1987 by Texas Instruments Incorporated
Published by John Wiley & Sons, Inc.
All rights reserved. Published simultaneously in Canada.

Reproduction or translation of any part of this work
beyond that permitted by Section 107 or 108 of the
1976 United States Copyright Act without the permission
of the copyright owner is unlawful. Requests for
permission or further information should be addressed to
the Permissions Department, John Wiley & Sons, Inc.

Library of Congress Cataloging-in-Publication Data:

Parks, T. W.
  Digital filter design.

  "A Wiley-Interscience publication."
  Includes bibliographies and index.
  1. Electric filters, Digital—Design and
construction. I. Burrus, C. S. II. Title.
TK7872.F5P37   1987   621.3815'324   86-32500
ISBN 0-471-82896-3

Printed in the United States of America

10 9 8 7 6 5 4 3 2 1

QUEEN MARY
COLLEGE
LIBRARY

To our parents
William and Mildred Parks
and
Aleta Huffman

Preface

This digital filter design book is addressed to the mathematician, scientist, or engineer who has an understanding of continuous-time signals and who has been introduced to discrete-time signal analysis.

The main topic of this book is the frequency-domain analysis and design of linear, constant coefficient, digital filters. The book is divided into two major parts: finite-duration impulse-response (FIR) filters and infinite-duration impulse-response (IIR) filters. Each part consists of a complete, self-contained treatment of the corresponding filter type. All aspects of each filter type are discussed. Each part begins with a discussion of filter properties, which leads into material on design of the filter to meet frequency-domain specifications. This aspect of filter design is called the *approximation problem* and makes up a major portion of the book. Each of the two parts concludes with a chapter on implementation of the filter with fixed-point arithmetic—the *realization problem*. The chapters on implementation both include a detailed design example that presents a step-by-step design and implementation of a typical filter. The design examples begin with the frequency-domain specifications for the filter and conclude with a listing of the assembly language program for implementing the filter on a signal-processing chip (the TMS32010 from Texas Instruments). The book begins with an introductory chapter that reviews the concepts of frequency-domain analysis of discrete-time systems and states the major problems in digital filter design. The final chapter summarizes the main results in the book with a discussion of the unique characteristics of the FIR and IIR filter types. An appendix with listings of ten FORTRAN programs for filter design is included.

This book may be used in several ways. For some applications one might turn to the appendix, run the appropriate design program to get the coefficients of a filter that meets given frequency-domain specifications, then turn to the listing in the design example, insert the coefficients in the listing, and run the program on a TMS32010. If all goes smoothly in this process, one may not need to read and completely understand the theory in the book. However, if, as often happens, the problem one is faced with is not exactly covered by the programs in the appendix, then with some reading of the theory, one can probably modify the appropriate design program or write a special program to obtain the appropriate filter coefficients. Even if the coefficients can be obtained from a program in the appendix, the implementation in the design example may not be exactly what one wants. For example, the filter may take too much time to execute or may require too much memory or may have undesirable quantization effects. Again, some time spent in reading the theory in the chapters on implementation should allow the reader to develop an appropriate implementation of the desired digital filter.

This book would not have been written without the support and encouragement of Texas Instruments, Inc. We would especially like to thank Mike Hames, who has always been ready with a smile and a helping hand when all of us realized just how much work is involved in writing a book. Maridene Lemmon has continued to patiently correct and improve our writing styles and has carefully read through countless revisions of the manuscript. The engineers at TI have read early versions of the text and helped correct our errors.

We would like to thank Professor H. W. Schüssler who helped us begin to understand the issues in digital filter implementation when he was on leave at Rice University. Some of our examples are taken from his notes. We would like also to thank Cole Erskine for working out the two detailed design examples and for providing the necessary TMS32010 code. Jim Kaiser and Dick Roberts provided us with very thorough reviews of the manuscript and made several good suggestions, which we have incorporated in the text.

We appreciate the long hours of reading put in by our graduate students Doug Jones and Henrik Sorenson, who have made many good suggestions for improving the book. Thanks also is given to the students at Rice University in our digital signal-processing courses who have helped us develop this book over the years.

T. W. PARKS
C. S. BURRUS

Contents

Examples

Digital Filter Design

Part I

Introduction

1

Introduction to Digital Filters

Digital filters, at first, were simulations of analog filters on general-purpose computers. As computer technology provided faster multipliers, more memory, and good analog-to-digital converters, some of these computer simulations were implemented in special hardware to replace the analog filter. These early digital filters were large, expensive, and consumed considerable power. Nevertheless, in applications where the flexibility and programmability of a digital filter could be put to good use, and where cost, power consumption, and volume were not major considerations, a digital approach to filtering was sometimes superior to conventional analog filtering.

In the late 1960s and early 1970s, at a series of IEEE-sponsored Arden House workshops, it seemed that there was always a talk comparing analog and digital filters. Leaders in the field expressed the opinion that although digital filters cost too much, were too large, and/or used too much power, they offered the following advantages:

1. Programmable (filter characteristics easily changed)
2. Reliable and repeatable
3. Free from component drift
4. No tuning required
5. No precision components, no component matching
6. Superior performance (linear phase, for example)

These talks comparing analog and digital filters would inevitably end on an optimistic note, predicting that in two years, at the next Arden House meeting, digital technology would have advanced enough so that digital filters would be the better choice for most applications in the audio-frequency range. These

3

predictions were made over a period of several years. While digital processing was becoming faster and less expensive, analog filter technology was also making advances.

In the 1980s, very large-scale integration (VLSI) developments have dramatically reduced the cost and power consumption of digital filters and have led to much more widespread application of digital signal processing. Digital filters are even finding their way into the home in compact disk players and television sets.

This book has four major parts. Part I (this chapter) contains an introduction and reviews the concepts of linear discrete-time systems, frequency response, and filtering. Descriptions of the approximation and realization problems for digital filters are also included along with a discussion of FIR (finite-duration impulse-response) and IIR (infinite duration impulse-response) designs. Parts II and III form the main body of the book. Part II, in Chapters 2 through 5, gives a complete treatment of the properties of FIR filters, the approximation problem for linear-phase, minimum-phase, and complex designs, and the implementation of FIR filters with fixed-point arithmetic. Part III, in Chapters 6 through 8, treats properties, design, and implementation of IIR filters. Each part concludes with a design example that gives a step-by-step design, beginning with the specifications and concluding with an implementation on a Texas Instruments TMS32010 signal-processing chip.

The design examples provide details of the design and implementation of typical low-pass filters, including assembly language programs for the TMS32010 digital signal processor. FORTRAN programs are supplied in the appendix for designing both FIR and IIR filters. Part IV contains a concluding chapter on the various merits of FIR and IIR designs and compares characteristics of these two filter types.

1.1 PROPERTIES OF DISCRETE-TIME SYSTEMS

A discrete-time system takes an input sequence of numbers and produces an output sequence of numbers. These number sequences are often samples of a continuous function of time, where $x(n)$ represents the signal $x(t)$ at equally spaced times $t_n = nT$—hence the name *discrete-time*. To simplify the discussion, we let $T = 1$, unless otherwise specified.

The box labeled S in Fig. 1.1 represents a discrete-time system with input x and output y. When the input $x_1(n)$ gives the output $y_1(n)$, and $x_2(n)$ gives $y_2(n)$ the system S in Fig. 1.1 is called *linear* if the linear combination of the inputs $a_1x_1(n) + a_2x_2(n)$ produces the output $a_1y_1(n) + a_2y_2(n)$ for any choice of constants a_1 and a_2.

The system in Fig. 1.1 is called *stationary* or *time invariant* when

$$x_1(n) \longrightarrow \boxed{\text{System } S} \longrightarrow y_1(n)$$

implies that

$$x_1(n-m) \longrightarrow \boxed{\text{System } S} \longrightarrow y_1(n-m)$$

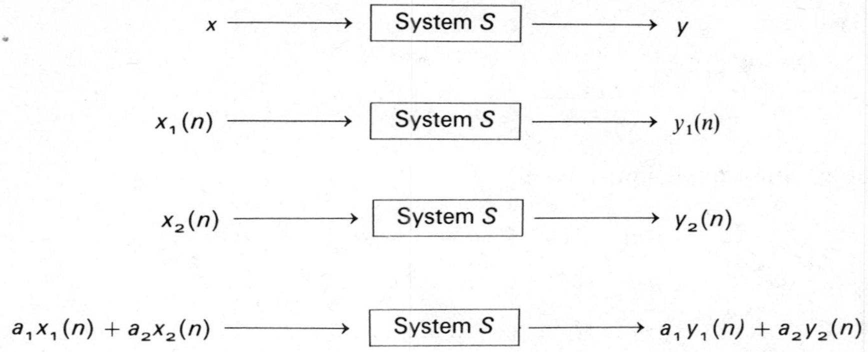

$$x \longrightarrow \boxed{\text{System } S} \longrightarrow y$$

$$x_1(n) \longrightarrow \boxed{\text{System } S} \longrightarrow y_1(n)$$

$$x_2(n) \longrightarrow \boxed{\text{System } S} \longrightarrow y_2(n)$$

$$a_1 x_1(n) + a_2 x_2(n) \longrightarrow \boxed{\text{System } S} \longrightarrow a_1 y_1(n) + a_2 y_2(n)$$

FIGURE 1.1. A discrete-time system.

for any time shift m. The system S is *stable* if the output $y(n)$ remains bounded for any bounded input signal $x(n)$. The discrete-time system S is called a *digital system* if the input $x(n)$ and output $y(n)$ can assume only a finite number of possible values or if $x(n)$ and $y(n)$ are quantized.

1.2 LINEAR, STATIONARY, DISCRETE-TIME SYSTEMS

The analysis of a linear system consists of the following three steps:

1. Resolution of the input into simple components
2. Calculation of the system response to these simple components
3. Superposition of the responses

For time-domain analysis of a discrete-time system, the simple components are shifted unit-sample functions $\delta(n - m)$, also called a digital impulse or unit-pulse signal, where the unit-sample function $\delta(n)$ is equal to unity at $n = 0$ and is equal to zero elsewhere, as shown in Fig. 1.2. The three steps in the analysis are as follows:

1. The input is resolved into shifted unit-sample functions, $\delta(n)$, weighted by the value of the signal where the unit-sample function is located, as indicated in Fig. 1.3.

$$x(n) = \sum_m x(m)\delta(n - m). \tag{1.1}$$

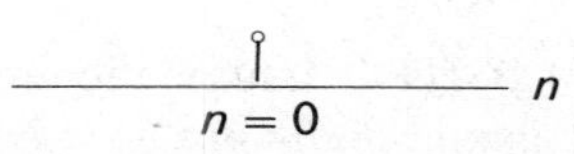

FIGURE 1.2. The unit-sample function.

Signal

$$m = 0 \qquad m = 1 \qquad m = 2$$

Shifted unit-sample functions

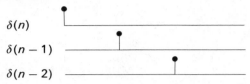

$\delta(n)$

$\delta(n - 1)$

$\delta(n - 2)$

FIGURE 1.3. Resolution of a signal into unit-sample functions.

2. The response of the system to a unit pulse located at the origin is called the unit-pulse response $h(n)$. When the system is stationary, the response to a shifted unit pulse $\delta(n - m)$ is simply the shifted unit-pulse response $h(n - m)$.

$$\delta(n) \longrightarrow \boxed{\text{System } S} \longrightarrow h(n)$$

$$\delta(n - m) \longrightarrow \boxed{\text{System } S} \longrightarrow h(n - m)$$

3. Superposition is used to add the individual responses.

$$y(n) = \sum_m x(m)h(n - m). \tag{1.2}$$

In (1.2), $h(n - m)$ is the system response to a unit-sample function located at time m, and $x(m)$ is the actual input value at time m. The summation of output components in (1.2) is called *discrete-time convolution* or simply *convolution*.

A *causal* system has a unit-pulse response that is zero for $n < 0$, and the convolution in (1.2) becomes

$$y(n) = \sum_{m = -\infty}^{n} x(m)h(n - m), \tag{1.3}$$

or, with a change of variable,

$$y(n) = \sum_{m = 0}^{\infty} h(m)x(n - m). \tag{1.4}$$

The summations in (1.3) and (1.4) assume the unit-pulse response has infinite duration; that is, the filter is an *infinite-duration impulse-response* (IIR) filter. If the unit-pulse response of a causal system is zero for all $n > N - 1$, the convolution in (1.4) becomes

$$y(n) = \sum_{m=0}^{N-1} h(m)x(n - m), \tag{1.5}$$

and the filter is called a *finite-duration impulse-response* (FIR) filter.

1.3 FREQUENCY RESPONSE AND TRANSFER FUNCTIONS

If the input to a causal, linear, stationary system is a complex exponential with frequency ω,

$$x(n) = e^{j\omega n}. \tag{1.6}$$

Then by (1.4)

$$y(n) = \sum_{m=0}^{\infty} h(m)e^{j\omega(n-m)} = e^{j\omega n}\left[\sum_{m=0}^{\infty} h(m)e^{-j\omega m}\right]. \tag{1.7}$$

When the summation over m on the right side of (1.7) is written

$$H(\omega) = \sum_{m=0}^{\infty} h(m)e^{-j\omega m}, \tag{1.8}$$

then the response $y(n)$ to an exponential input at frequency ω is

$$y(n) = H(\omega)e^{j\omega n}. \tag{1.9}$$

Thus, $H(\omega)$ describes the change in magnitude and phase at the frequency ω— hence the name *frequency response*.

If the input $x(n)$ is a real-valued signal $2\cos(\omega n)$, it can be written as the sum of two complex exponentials:

$$x(n) = e^{j\omega n} + e^{-j\omega n}, \tag{1.10}$$

and the output is

$$y(n) = H(\omega)e^{j\omega n} + H^*(\omega)e^{-j\omega n}, \tag{1.11}$$

where $H^*(\omega)$ is the complex conjugate of $H(\omega)$.

Letting $|H(\omega)|$ and $\theta(\omega)$ be the magnitude and phase of $H(\omega)$, respectively, we can write

$$y(n) = 2|H(\omega)|\cos(\omega n + \theta(\omega)). \tag{1.12}$$

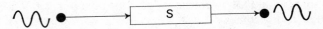

FIGURE 1.4. Frequency response of a linear system.

As indicated in Fig. 1.4, the input cosine signal experiences an amplitude change according to the magnitude $|H(\omega)|$ of the frequency response and a phase shift according to the phase $\theta(\omega)$ of the frequency response. If the phase-shift term is rewritten as

$$\tau_p = \frac{-\theta(\omega)}{\omega}, \qquad (1.13)$$

then the output signal $y(n)$ experiences a delay of τ_p. Thus, τ_p is called the *phase delay*[1,2] of the system.

$$y(n) = 2|H(\omega)| \cos(\omega(n - \tau_p)). \qquad (1.14)$$

The *group delay* of a system is defined to be[1,2]

$$\tau_g = -\frac{d\theta(\omega)}{d\omega}. \qquad (1.15)$$

For a bandpass signal the phase delay τ_p represents the delay of the carrier, and the group delay τ_g represents the delay of the envelope of the signal.

In contrast to a continuous-time system, the frequency response of a discrete-time system is always periodic with a period equal to the sampling frequency, which in this case is normalized to one sample per second or 2π radians per second. This is shown by

$$H(\omega + 2\pi) = \sum_{m=0}^{\infty} h(m)e^{-j(\omega + 2\pi)m} = H(\omega). \qquad (1.16)$$

Another useful property of the frequency response of a system with a real-valued unit-pulse response is

$$H^*(\omega) = H(-\omega).$$

This relation means that the frequency response has an even magnitude function and an odd phase function. Thus, for systems with real-valued unit-pulse responses, frequency-response plots need only be drawn for positive frequencies.

The response shown in (1.7) assumes that the input signal was an exponential with frequency ω for all time n. If the input signal begins or ends at a defined time, (1.7) does not apply. The response of the system is then the sum of a steady-state component, determined from the frequency response, and a transient component. This distinction becomes especially important for short-duration inputs. If, for example, an interfering tone is to be filtered out, a filter with a zero response at the frequency of the tone is designed. If the tone starts and then stops after a short time, the response of the filter will not be zero, as it would be if the tone had been present for all time. Only the steady-state component of the response will be zero. Transient, nonzero output components are produced when the tone begins and ends.

We can generalize the idea of representing the behavior of a system in terms of its frequency response by using the z transform.[1-4] If we use a complex number z, written

$$x(n) = z^n, \tag{1.17}$$

in the convolution sum of (1.4), instead of an exponential input, as in (1.6), the output is

$$y(n) = H(z)z^n, \tag{1.18}$$

where

$$H(z) = \sum_{n=0}^{\infty} h(n)z^{-n} \tag{1.19}$$

is the **z-transform** of the unit-pulse response $h(n)$. The z transform of the unit-pulse response is also called the **transfer function** of the system. For the summation in (1.19) to converge, the magnitude of z must be large enough, or

$$|z| > R. \tag{1.20}$$

If the region in the complex z plane, given by (1.20), where (1.19) converges includes the unit circle (i.e., if $R \leqslant 1$), then the transfer function in (1.19), when evaluated on the unit circle, is simply the frequency response of the system given in (1.18).

$$H(z)\bigg|_{z=e^{j\omega}} = \sum_{n=0}^{\infty} h(n)e^{-j\omega n} = H(e^{j\omega}). \tag{1.21}$$

The use of the same function H for both the transfer function and the frequency response is quite common in the literature and should not cause confusion when taken in context. It is easier to write $H(\omega)$ rather than $H(e^{j\omega})$ for the frequency response.

1.4 DIGITAL FILTER DESIGN

The two parts to the filter design process are the approximation problem and the realization problem. The approximation part of the problem deals with the choice of parameters or coefficients in the filter's transfer function to approximate an ideal or desired response. This approximation is often performed by using the frequency response.

The realization part of the filter design problem deals with choosing a structure to implement the transfer function. This structure may be in the form of a circuit diagram if the filter is to be built of components, or it may be a program to be used on a general-purpose computer or a signal-processing microprocessor.

The approximation stage takes the specification and gives a transfer function through four steps:

1. A desired or ideal response is chosen, usually in the frequency domain.
2. An allowed class of filters is chosen (e.g., length-N FIR filters).
3. A measure of the quality of the approximation is chosen (e.g., maximum error in the frequency domain).
4. A method or algorithm is selected to find the best filter transfer function.

The realization stage then takes this transfer function and gives a circuit or program through four steps:

1. An allowed set of structures is chosen.
2. A measure of the performance of the structure is chosen (e.g., the minimization of quantization noise).
3. The best structure is chosen from the allowed set, and its parameters are calculated from the transfer function.
4. The structure is implemented as a circuit or as a program.

These steps in filter design are not independent of each other; therefore, some iteration is often required. However, to do the best job of filter design, one must recognize and understand these distinct steps.

1.4.1 The Approximation Problem

Recall from Section 1.3 that the transfer function is defined as the z transform of the unit-pulse response of the filter.[1-4] The digital filters in this book are all assumed to be causal and can all be characterized by a transfer function

$$H(z) = \frac{b_0 + b_1 z^{-1} + \cdots + b_M z^{-M}}{1 + a_1 z^{-1} + \cdots + a_N z^{-N}}. \tag{1.22}$$

The region of convergence for $H(z)$ lies outside a circle centered at the origin of the z plane. This circle passes through the pole with the largest radius. For stable filters this radius is less than unity. If the transfer function of a filter can be written as a polynomial (all $a_i = 0$), the filter has a finite-duration unit-pulse response and is called an FIR filter. If common factors are allowed in the numerator and denominator of (1.22), the transfer function of an FIR filter can be written as a rational function with some $a_i \neq 0$. If, after cancellation of all factors common to both the numerator and denominator, some of the a_i coefficients in the denominator are not equal to zero, then the filter has an infinite-duration unit-pulse response and is called an IIR filter.

The approximation problem for an FIR filter is usually stated in the frequency domain, equivalent to $z = e^{j\omega}$. The filter parameters to be chosen are the unit-pulse response values b_i in

$$H(e^{j\omega}) = b_0 + b_1 e^{-j\omega} + \cdots + b_M e^{-j\omega M}, \tag{1.23}$$

the Mth-order numerator polynomial of (1.22) being evaluated on the unit circle. More conventional terminology refers not to the order of the FIR filter but to its length. A length-N FIR filter has a frequency response

$$H(e^{j\omega}) = h_0 + h_1 e^{-j\omega} + \cdots + h_{N-1} e^{-j\omega(N-1)}, \tag{1.24}$$

with N unit-pulse response values h_i, $i = 0, \ldots, N - 1$.

In its most general form the approximation problem is a polynomial (for FIR filters) or rational (for IIR filters) approximation problem with a complex desired function on the frequency band from $-\pi$ to π. (Recall that a digital filter has a periodic frequency response with a period of 2π.) The parameters a_i and b_i are chosen to minimize an appropriate metric of the distance between the desired response $D(z)$ and the actual response $H(z)$, often the norm of the difference,

$$\|E(z)\| = \left\| D(z) - \frac{b_0 + b_1 z^{-1} + \cdots + b_M z^{-M}}{1 + a_1 z^{-1} + \cdots + a_N z^{-N}} \right\|. \tag{1.25}$$

The complex, nonlinear problem that results with IIR design has no completely satisfactory solution to date. The much simpler complex, linear problem for FIR filters can be solved by linear programming, as described in Chapter 4. Usually the approximation problem is stated as a real approximation problem where the squared magnitude of the frequency response, a real-valued function, is chosen to meet a specification on the magnitude squared, and the filter is forced to be a minimum-phase filter (see Chapter 4). For FIR filters with exactly linear phase a real approximation problem also results.

Several possible choices are available for the norm function in (1.25). The most widely used are the Chebyshev norm and the least square or l_2 norm. The Chebyshev norm is appropriate when specifications are stated in terms of minimum allowed stop-band attenuation or maximum allowed pass-band error. The least squared error measure is appropriate when specifications are in terms of signal energy.

1.4.2 The Realization Problem

After the coefficients in the transfer function have been chosen, the problem is only partially solved. The second part of the filter design is the realization problem. Choices must be made concerning methods for implementing the filter. The transfer function

$$H(z) = \frac{b_0 + b_1 z^{-1} + \cdots + b_M z^{-M}}{1 + a_1 z^{-1} + \cdots + a_N z^{-N}} \tag{1.26}$$

corresponds to a difference equation relating the output $y(n)$ and the input $x(n)$.

$$y(n) = b_0 x(n) + b_1 x(n-1) + \cdots + b_M x(n-M)$$
$$- a_1 y(n-1) - a_2 y(n-2) - \cdots - a_N y(n-N). \tag{1.27}$$

One implementation or realization of the filter is the direct calculation implied by (1.27). As shown in Chapter 8, this direct method may not be satisfactory when the coefficients are quantized. In fact, it is this quantization of coefficients and signal values that makes the choice of filter realization important. Just as there are different methods for evaluating a polynomial (e.g., Horner's rule), there are many different ways to calculate the output $y(n)$ in (1.27). These different methods for calculating filter outputs may be conveniently represented by block diagrams that illustrate alternative filter structures. For example, the two diagrams in Figs. 1.5 and 1.6 represent the calculation of the same difference equation describing the input/output relation for a second-order IIR filter. However, these two structures have different properties when the coefficients and signal values are quantized. The boxes with z^{-1} represent delay elements, and the coefficients a_i and b_i are represented as gains on the various branches of

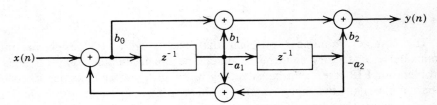

FIGURE 1.5. Direct-form second-order block.

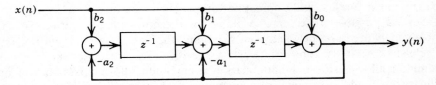

FIGURE 1.6. Transpose-form second-order block.

the diagrams. Finite word-length effects, such as overflow, quantization noise, and coefficient errors caused by quantized coefficients, demand an understanding of the possible structures for implementing a digital filter.

The realization chosen for a specific filter is also based on the type of digital hardware available for the implementation. With custom hardware, 6- or 7-bit coefficients may be used to save hardware, and some registers may have 8 bits, 12 bits, and so on. Parallel computations may be the best implementation of the filter. Input/output of the signal may be an important, time-consuming operation, or the multiplications may take the longest time. In the case of a programmable microprocessor, the instruction set plays an important role in selecting a realization for a filter. For example, the TMS320 family has special instructions to facilitate the multiply/accumulate operation in the direct realization of FIR filters.

1.5 PROPERTIES OF FIR AND IIR FILTERS

The two types of filters, FIR and IIR, are treated in separate parts of this book. They have very different characteristics, yet they can often meet the same specifications. The FIR filter has a transfer function that is a polynomial in z^{-1} and is an all-zero filter in the sense that the zeros in the z plane determine the frequency-response magnitude characteristic. Although a length-N FIR filter has a pole of order $N - 1$ at the origin of the z plane, a pole at the origin does not affect the magnitude of the frequency response of the filter. An FIR filter can have a unit-pulse response that is symmetric around the point $(N - 1)/2$ and can therefore have exactly linear phase.

The IIR filter has both poles and zeros in the z^{-1} plane and in the z plane. The combination of a pole near the pass-band edge with a zero near the stop-band edge can give an IIR filter a very short transition region between the pass-band and the stop band. Generally, an IIR filter can give a sharper cutoff than an FIR filter of the same order because both poles and zeros are present. However, a causal IIR filter cannot achieve exactly linear phase but the FIR filter can. The phase and group delay characteristics of conventional IIR filters are generally not as good as those of FIR filters.

Which filter is better for a particular application depends on the hardware used for the implementation of the filter. For example, the TMS320 family of signal processors has special instructions to facilitate the implementation of an FIR filter, a length-N FIR filter can be computed in about the same time as an IIR filter of order $N/2.5$ for the TMS32010 and $N/5.0$ for the TMS32020.[9] On the other hand, the IIR filter requires less memory than an FIR filter that meets about the same frequency-domain specifications. The IIR filter, when implemented in fixed-point arithmetic, may have instabilities (limit cycles) and may have large quantization noise, depending on the number of bits allocated to the coefficients and the signal variables in the filter. The FIR filter, on the other hand, is usually implemented in a nonrecursive way, which guarantees a stable filter.

For narrow-band, sharp cutoff filters where phase is not important, IIR filters are likely to be superior to FIR filters. For applications where wave shape is important, the FIR filter with its good phase characteristics is usually a good choice.

REFERENCES

Considerable literature is available on the subject of analog and digital filter design. Some books that give good descriptions of the theory of filter design are listed in the references which follow.

[1]   A. V. Oppenhim and R. W. Schafer, *Digital Signal Processing*, Englewood Cliffs, N.J.: Prentice-Hall, 1975.

[2]   L. R. Rabiner and B. Gold, *Theory and Application of Digital Signal Processing*, Englewood Cliffs, NJ: Prentice-Hall, 1975.

[3]   F. J. Taylor, *Digital Filter Design Handbook*, New York: Dekker, 1983.

[4]   R. A. Roberts and C. T. Mullis, *Digital Signal Processing*, Reading, MA: Addison-Wesley, 1987.

[5]   L. R. Rabiner and C. M. Rader, eds., *Digital Signal Processing*, selected reprints, New York: IEEE Press, 1972.

[6]   *Digital Signal Processing II*, selected reprints, New York: IEEE Press, 1979. Edited by the Digital Signal Processing Committee.

[7]   B. Gold and C. M. Rader, *Digital Processing of Signals*, New York: McGraw-Hill, 1969.

[8]   *Programs for Digital Signal Processing*, New York: IEEE Press, 1979.

[9]   *Digital Signal Processing Applications with the TMS320 Family*, Texas Instruments, 1986.

Part II

Finite Impulse-Response
(FIR) Filters

2

Properties of Finite Impulse-Response Filters

Digital filters with a finite-duration impulse-response (FIR) have characteristics that make them useful in many applications.[1,2] They can achieve exactly linear phase and cannot be unstable. The design methods are generally linear. They can be efficiently realized on general- or special-purpose hardware. This chapter examines and evaluates important design characteristics of the four basic types of linear-phase FIR filters.

Because of the method of implementation, the FIR filter is also called a nonrecursive filter or a convolution filter. From the time-domain view of this operation, the FIR filter is sometimes called a moving-average filter. All of these names represent useful interpretations that are discussed in this chapter; however, the name "FIR" is the one most commonly seen in filter design literature.

The duration or sequence length of the impulse response of these filters is, by definition, finite; therefore, the output can be written as a finite convolution sum

$$y(n) = \sum_{m=0}^{N-1} h(m)x(n-m). \tag{2.1}$$

With a change of index variables, we can also write

$$y(n) = \sum_{m=n}^{n-N+1} h(n-m)x(m), \tag{2.2}$$

where $x(n)$ is the input and $h(n)$ is the length-N impulse response.

The FIR filter may be interpreted as an extension of a moving sum or as a weighted moving average. If one has a sequence of numbers, such as prices from the daily stock market for a particular stock, and would like to remove the erratic variations in order to discover longer trends, each number could be replaced by the average of itself and the preceding three numbers; that is, the variations within a four-day period would be "averaged out," and the longer-term variations would remain. To illustrate how this happens, we consider an artificial signal $x(n)$ containing a linear term, $K_1 n$, and an undesired oscillating term added to it, such that

$$x(n) = K_1 n + K_2 \cos(\pi n). \tag{2.3}$$

If a length-2 averaging filter is used with

$$h(n) = \begin{cases} \frac{1}{2}, & n = 0, 1, \\ 0, & \text{otherwise.} \end{cases}$$

it can be verified that, after two outputs, the output is exactly the linear term with a delay of one-half sample interval and no oscillation.

This example illustrates the basic FIR filter-design problem: determine N, the number of terms for $h(n)$, and the values of $h(n)$ for achieving a desired effect on the signal. Simple examples should be attempted to obtain an intuitive idea of the FIR filter as a moving average; however, this approach will not suffice for complicated problems where the concept of frequency becomes more valuable.

2.1 FREQUENCY-DOMAIN DESCRIPTION OF FIR FILTERS

The transfer function of an FIR filter is given by the z transform of $h(n)$ as

$$H(z) = \sum_{n=0}^{N-1} h(n) z^{-n}. \tag{2.4}$$

The frequency response of a filter, as shown in Section 1.3, is found by setting $z = e^{j\omega}$, which gives (2.4) the form

$$H(\omega) = \sum_{n=0}^{N-1} h(n) e^{-j\omega n}, \tag{2.5}$$

where ω is frequency in radians per second. Strictly speaking, the exponent should be $-j\omega Tn$, where T is the time interval between the integer steps of n (the sampling interval). We assume that $T = 1$, until later in the book where the relation between n and time is important. To simplify notation, we let $H(\omega)$ rather than $H(e^{j\omega})$ represent the frequency response. It should always be clear from the context whether H is a function of z or ω.

This frequency-response function is complex valued and consists of a magnitude and a phase. Even though the impulse response is a function of the discrete variable n, the response is a function of the continuous-frequency variable ω and is periodic with period 2π. This periodicity is easily shown as follows:

$$H(\omega + 2\pi) = \sum_{n=0}^{N-1} h(n)e^{-j(\omega + 2\pi)n} \qquad (2.6)$$

$$= \sum_{n=0}^{N-1} h(n)e^{-j\omega n}e^{-j2\pi n}$$

$$= H(\omega).$$

Frequency is denoted by ω in radians per second or by f in hertz (cycles per second). These quantities are related by

$$\omega = 2\pi f.$$

An example of a length-5 filter is $h(n) = \{2, 3, 4, 3, 2\}$. Figure 2.1 shows the frequency-response plot.

The discrete Fourier transform (DFT) can be used to evaluate the frequency response at certain frequencies. The DFT[3] of the length-N impulse response $h(n)$

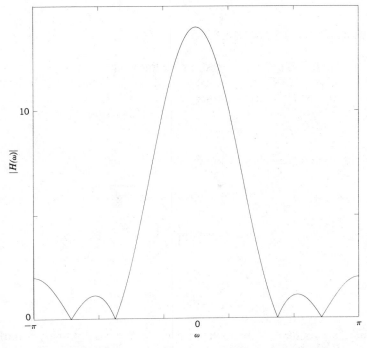

FIGURE 2.1. Frequency response of the example $h(n)$.

is defined as

$$C(k) = \sum_{n=0}^{N-1} h(n)e^{-j2\pi nk/N}, \qquad k = 0, 1, \ldots, N-1, \tag{2.7}$$

When compared to (2.5), (2.7) gives

$$C(k) = H(\omega)\bigg|_{\omega = 2\pi k/N} = H\left(\frac{2\pi k}{N}\right), \qquad k = 0, 1, \ldots, N-1. \tag{2.8}$$

This states that the DFT of $h(n)$ gives N samples of the frequency-response function $H(\omega)$. This sampling at N points may not give enough detail; therefore, more samples are needed. Any number of equally spaced samples can be found with the DFT by simply appending $L - N$ zeros to $h(n)$ and taking an L-length DFT. This method is often useful when an accurate picture of all of $H(\omega)$ is required. Indeed, when the number of appended zeros goes to infinity, the DFT becomes the Fourier transform of $h(n)$.

Because the DFT of $h(n)$ is a set of N samples of the frequency response FIR filters can be designed so that the inverse DFT of N samples of a desired frequency response gives the filter coefficients $h(n)$. That approach is called frequency sampling and is developed in Section 3.1.

2.2 LINEAR-PHASE FIR FILTERS

Particularly useful FIR filters are those with linear phase shift.[1,2] To develop the theory for this set of FIR filters, we need a definition of phase shift. If the real and imaginary parts of $H(\omega)$ are given by

$$H(\omega) = R(\omega) + jI(\omega),$$

the magnitude and phase are defined by

$$M(\omega) = |H(\omega)| = \sqrt{R^2 + I^2},$$

$$d(\omega) = \arctan\left(\frac{I}{R}\right).$$

Thus

$$H(\omega) = M(\omega)e^{jd(\omega)}. \tag{2.9}$$

Mathematical problems arise, however, because $M(\omega)$ is not analytic and $d(\omega)$ is not continuous. This problem is solved by introducing the real-valued amplitude function $A(\omega)$ that may be positive or negative. The frequency response is written as

$$H(\omega) = A(\omega)e^{j\theta(\omega)}, \tag{2.10}$$

where $A(\omega)$ is called the amplitude to distinguish it from the magnitude $M(\omega)$, and $\theta(\omega)$ is the continuous version of $d(\omega)$. $A(\omega)$ is a real, analytic function related to the magnitude by

$$A(\omega) = \pm M(\omega) \tag{2.11}$$

or $|A(\omega)| = M(\omega)$. With this definition, $A(\omega)$ can be made analytic and $\theta(\omega)$ can be made continuous. These quantities are much easier to work with than $M(\omega)$ and $d(\omega)$. The relationships of $A(\omega)$ and $M(\omega)$ and of $\theta(\omega)$ and $d(\omega)$ are shown in Fig. 2.2.

To develop the characteristics and properties of linear-phase filters, we

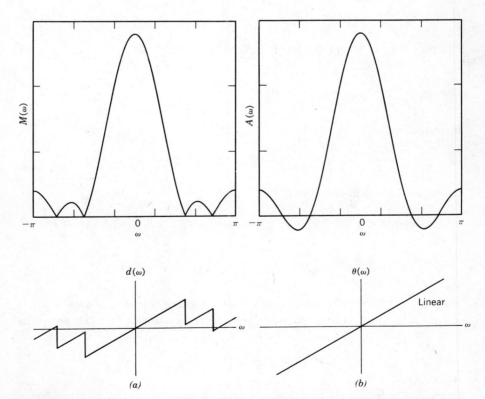

FIGURE 2.2. The magnitude and amplitude of an example linear-phase FIR filter. (a) magnitude and phase. (b) amplitude and phase.

assume a general linear form for the phase function:

$$\theta(\omega) = K_1 + K_2\omega. \tag{2.12}$$

Equation (2.5) gives the frequency-response function of a length-N FIR filter as

$$H(\omega) = \sum_{n=0}^{N-1} h(n)e^{-j\omega n}, \tag{2.13}$$

$$H(\omega) = e^{-j\omega M} \sum_{n=0}^{N-1} h(n)e^{j\omega(M-n)},$$

and

$$H(\omega) = e^{-j\omega M}[h_0 e^{j\omega M} + h_1 e^{j\omega(M-1)} + \cdots + h_{N-1}e^{j\omega(M-N+1)}]. \tag{2.14}$$

Equation (2.14) can be written in the form

$$H(\omega) = A(\omega)e^{j(K_1 + K_2\omega)} \tag{2.15}$$

if M (not necessarily an integer) is defined by

$$M = \frac{N-1}{2}, \tag{2.16}$$

or, equivalently,

$$M = N - M - 1.$$

Equation (2.14) then becomes

$$\begin{aligned} H(\omega) = e^{-j\omega M}\{&(h_0 + h_{N-1})\cos(\omega M) + j(h_0 - h_{N-1})\sin(\omega M) \\ &+ (h_1 + h_{N-2})\cos(\omega(M-1)) \\ &+ j(h_1 - h_{N-2})\sin(\omega(M-1)) + \cdots\}. \end{aligned} \tag{2.17}$$

We can put (2.17) in the form of (2.15), where $A(\omega)$ is real, in two ways: $K_1 = 0$ or $K_1 = \pi/2$. The first case requires a special even symmetry in $h(n)$ of the form

$$h(n) = h(N - n - 1), \tag{2.18}$$

which gives

$$H(\omega) = A(\omega)e^{-jM\omega},$$

where $A(\omega)$ is a real-valued function of ω and $e^{-jM\omega}$ gives the linear phase. When N is odd,

$$A(\omega) = \sum_{n=0}^{M-1} 2h(n)\cos(\omega(M-n)) + h(M). \tag{2.19}$$

With a change of variables, we get

$$A(\omega) = \sum_{n=1}^{M} 2h(M-n)\cos(\omega n) + h(M). \tag{2.20}$$

When N is even,

$$A(\omega) = \sum_{n=0}^{N/2-1} 2h(n)\cos(\omega(M-n)). \tag{2.21}$$

With a change of variables, we get

$$A(\omega) = \sum_{n=1}^{N/2} 2h\left(\frac{N}{2}-n\right)\cos(\omega(n-\tfrac{1}{2})) \tag{2.22}$$

When $K_1 = \pi/2$ in (2.15), an odd symmetry of the form

$$h(n) = -h(N-n-1), \tag{2.23}$$

is required. For N odd, $H(\omega)$ then becomes

$$H(\omega) = jA(\omega)e^{-jM\omega},$$

where

$$A(\omega) = \sum_{n=0}^{M-1} 2h(n)\sin(\omega(M-n)), \tag{2.24a}$$

For N even,

$$A(\omega) = \sum_{n=0}^{N/2-1} 2h(n)\sin(\omega(M-n)). \tag{2.24b}$$

2.2.1 Four Types of Linear-Phase FIR Filters

From the previous discussion, we see that there are four possible types of FIR filters[1] leading to the linear phase of (2.15). We will consider each type.

Type 1. The impulse response has odd length and is even symmetric about its midpoint $n = M = (N - 1)/2$, which requires $h(n) = h(N - n - 1)$ and gives (2.19) and (2.20).

Type 2. The impulse response has even length and is even symmetric about M, but M is not an integer. Therefore, there is no $h(n)$ at the point of symmetry, but it satisfies (2.21) and (2.22).

Type 3. The impulse response has odd length and the odd symmetry of (2.23), giving an imaginary multiplier for the linear-phase form in (2.24a).

Type 4. The impulse response has even length and the odd symmetry of type 3 in (2.23) and (2.24b).

Examples of the four types of linear-phase FIR filters with the symmetries for odd and even lengths are shown in Fig. 2.3. Note that for N odd and antisymmetry, $h(M) = 0$.

To analyze or design linear-phase FIR filters, we need to know the characteristics of $A(\omega)$. The most important characteristics are shown in Table 2.1.

Figure 2.4 shows examples of amplitude functions for odd- and even-length linear-phase filters $A(\omega)$.

These characteristics reveal several inherent features that are extremely important to filter design. For types 3 and 4, $A(0) = 0$ for any choice of filter coefficients $h(n)$, which is undesirable for a low-pass filter. Types 2 and 3 always have $A(\pi) = 0$, which is undesirable for a high-pass filter. In addition to the linear-phase characteristic representing a time shift, types 3 and 4 give a constant 90° phase shift, desirable for a differentiator or a Hilbert transformer.

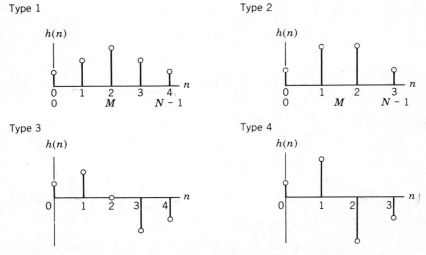

FIGURE 2.3. Example of impulse responses for the four types of linear-phase FIR filters.

TABLE 2.1. Characteristics of A(ω) for Linear Phase

Type 1. Odd length, even symmetric $h(n)$
$\quad A(\omega)$ is even about $\omega = 0$ $\qquad\qquad\qquad A(\omega) = A(-\omega)$
$\quad A(\omega)$ is even about $\omega = \pi$ $\qquad\qquad\quad A(\pi + \omega) = A(\pi - \omega)$
$\quad A(\omega)$ is periodic with period 2π $\qquad\quad A(\omega + 2\pi) = A(\omega)$

Type 2. Even length, even symmetric $h(n)$
$\quad A(\omega)$ is even about $\omega = 0$ $\qquad\qquad\qquad A(\omega) = A(-\omega)$
$\quad A(\omega)$ is odd about $\omega = \pi$ $\qquad\qquad\quad\; A(\pi + \omega) = -A(\pi - \omega)$
$\quad A(\omega)$ is periodic with period 4π $\qquad\quad A(\omega + 4\pi) = A(\omega)$

Type 3. Odd length, odd symmetric $h(n)$
$\quad A(\omega)$ is odd about $\omega = 0$ $\qquad\qquad\qquad A(\omega) = -A(-\omega)$
$\quad A(\omega)$ is odd about $\omega = \pi$ $\qquad\qquad\quad\; A(\pi + \omega) = -A(\pi - \omega)$
$\quad A(\omega)$ is periodic with period 2π $\qquad\quad A(\omega + 2\pi) = A(\omega)$

Type 4. Even length, odd symmetric $h(n)$
$\quad A(\omega)$ is odd about $\omega = 0$ $\qquad\qquad\qquad A(\omega) = -A(-\omega)$
$\quad A(\omega)$ is even about $\omega = \pi$ $\qquad\qquad\quad A(\pi + \omega) = A(\pi - \omega)$
$\quad A(\omega)$ is periodic with period 4π $\qquad\quad A(\omega + 4\pi) = A(\omega)$

Type 1

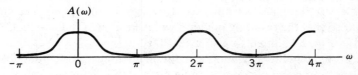

Type 2

Type 3

Type 4

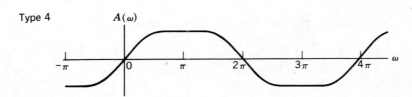

FIGURE 2.4. Examples of amplitude functions for FIR filters.

The first step in designing a linear-phase FIR filter is choosing the type most compatible with the specifications.

2.2.2 Calculation of FIR Filter Frequency Response

As shown in Section 2.1, L equally spaced samples of $H(\omega)$ are easily calculated for $L > N$ by appending $L - N$ zeros to $h(n)$ for a length-L DFT. This appears as

$$H\left(\frac{2\pi k}{L}\right) = \text{DFT}\{h(n)\} \quad \text{for } k = 0, 1, \ldots, L - 1. \tag{2.25}$$

This method is a straightforward and flexible approach. Only the samples of $H(\omega)$ that are of interest need to be calculated. In fact, we can even achieve nonuniform spacing of the frequency samples by altering the DFT defined in (2.7). Direct use of the DFT can be inefficient, and for linear-phase filters, it is $A(\omega)$, not $H(\omega)$, that is the most informative. In addition to direct application of the DFT, special formulas are developed in (2.26)–(2.29) for evaluating samples of $A(\omega)$ that exploit the fact that $h(n)$ is real and has certain symmetries. For long filters even these formulas are too inefficient, so the DFT is used but implemented by a fast Fourier transform (FFT) algorithm[3].

The DFT is first used to give $A(\omega)$ directly. If $h(n)$ is shifted to be symmetric about $n = 0$, the phase shift is zero; therefore $H(\omega) = A(\omega)$. The shift must be circular so that the resulting function will have a real DFT. Figure 2.5 shows the signal in Fig. 2.3 shifted to give a real DFT.

Because the point of symmetry is not on an integer, it is impossible to shift an even-length impulse response to give a real DFT. But we can circumvent this limitation by stretching the even-length signal to twice its length by placing a zero between each original value. The point of symmetry of this double-length signal then will be on an integer, and its DFT will be samples of two periods of the $A(\omega)$ of the original signal. This stretching and shifting is explained in reference 3 and illustrated in Fig. 2.6.

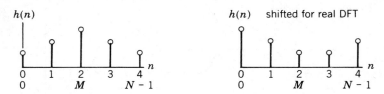

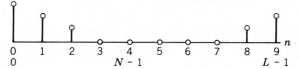

FIGURE 2.5. Shifted impulse response for real DFT.

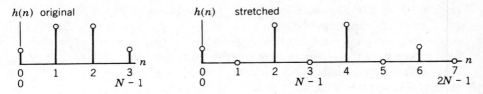

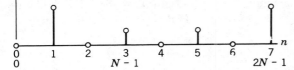

FIGURE 2.6. Modified even-length signal for a real DFT.

These DFT methods for calculating samples of $A(\omega)$ are inefficient because they do not take advantage of the symmetries and realness of $h(n)$. We can derive special formulas by building these characteristics into the DFT; see (2.19), (2.21), and (2.24), which evaluate $A(\omega)$ for any value of ω.

In the special case of type 1 filters with L equally spaced sample points, the samples of the frequency response have the form

$$A_k = A\left(\frac{2\pi k}{L}\right) = \sum_{n=0}^{M-1} 2h(n)\cos\left(\frac{2\pi(M-n)k}{L}\right) + h(M). \qquad (2.26)$$

For type 2 filters

$$A_k = A\left(\frac{2\pi k}{L}\right) = \sum_{n=0}^{N/2-1} 2h(n)\cos\left(\frac{2\pi(M-n)k}{L}\right). \qquad (2.27)$$

For type 3 filters

$$A_k = A\left(\frac{2\pi k}{L}\right) = \sum_{n=0}^{M-1} 2h(n)\sin\left(\frac{2\pi(M-n)k}{L}\right). \qquad (2.28)$$

For type 4 filters

$$A_k = A\left(\frac{2\pi k}{L}\right) = \sum_{n=0}^{N/2-1} 2h(n)\sin\left(\frac{2\pi(M-n)k}{L}\right). \qquad (2.29)$$

In all cases the midpoint or point of symmetry is $M = (N-1)/2$, which can be viewed as an average time delay for the filter. For the even-length filter this delay is not an integer multiple of the sample interval but gives a "half-sample delay". Formulas (2.26)–(2.29) are efficient methods for calculating the frequency response of an FIR filter with lengths up to a few hundred. The N^2 calculations required by this approach become too slow for longer lengths. A FORTRAN subroutine that implements both (2.26) and (2.27) is given as part of the low-pass filter programs in the appendix. A subroutine that implements (2.28) and (2.29) is in the differentiator design program in the appendix. These programs can easily be modified to drive a graphics terminal or plotter.

Although this section has primarily concentrated on linear-phase filters by taking their symmetries into account, the method of taking the DFT of $h(n)$ to obtain samples of the frequency response of an FIR filter also holds for general arbitrary phase filters.

2.2.3 Zero Locations for Linear-Phase FIR Filters

One can get a qualitative understanding of the filter characteristics by examining the locations of the $N - 1$ zeros of an FIR filter's transfer function. This transfer function is given by the z transform of the length-N impulse response from (2.4).

$$H(z) = \sum_{n=0}^{N-1} h(n)z^{-n}. \tag{2.30}$$

This equation can be rewritten as

$$H(z) = z^{-N+1}(h_0 z^{N-1} + h_1 z^{N-2} + \cdots + h_{N-1})$$

or

$$H(z) = z^{-N+1} D(z), \tag{2.31}$$

where $D(z)$ is an $(N - 1)$th-order polynomial that is multiplied by an $(N - 1)$th-order pole located at the origin of the complex z plane.

$h(n)$ real implies that the zeros will be real or will occur in complex conjugate pairs. If the FIR filter is linear phase, there are further restrictions on the possible zero locations. From (2.18) we see that linear phase implies a symmetry in the impulse response and, therefore, in the coefficients of the polynomial $D(z)$ in (2.31). Express the complex zero z_1 in polar form by

$$z_1 = r_1 e^{jx}, \tag{2.32}$$

where r_1 is the radial distance of z_1 from the origin in the complex z plane, and x is the angle from the real axis. See Fig. 2.7.

Using the definition of $H(z)$ and $D(z)$ in (2.30) and (2.31) and the linear-phase even-symmetry requirement of

$$h(n) = h(N - 1 - n)$$

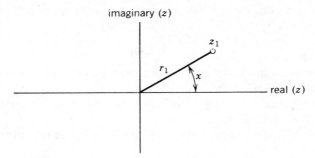

FIGURE 2.7. Location of the zero at z_1 in the complex plane.

give

$$H\left(\frac{1}{z}\right) = D(z). \tag{2.33}$$

Equation (2.33) implies that if z_1 is a zero of $H(z)$, then $1/z_1$ is also a zero of $H(z)$. In other words,

$$\text{if} \quad H(z_1) = 0, \quad \text{then} \quad H\left(\frac{1}{z_1}\right) = 0. \tag{2.34}$$

Equation (2.34) says that if a zero exists at a radius of r_1, then one also exists at a radius of $1/r_1$, thus giving a special type of symmetry of the zeros about the unit circle. Another possibility is that the zero lies on the unit circle with $r_1 = 1/r_1 = 1$.

There are four essentially different cases[1] of even-symmetric filters that have the lowest possible order. All higher-order symmetric filters have transfer functions that can be factored into products of these lowest-order transfer functions. They are illustrated by four basic filters of lowest order that satisfy these conditions: one length-2, two length-3, and one length-5.

The only length-2, even-symmetric, linear-phase FIR filter has the form

$$D(z) = (z + 1)K, \tag{2.35}$$

which, for any constant K, has a single zero at $z_1 = -1$.

The even-symmetric length-3 filter has a form

$$D(z) = (z^2 + az + 1)K. \tag{2.36}$$

There are two possible cases. For $|a| > 2$ two real zeros can satisfy (2.34) with $z_1 = r$ and $1/r$. Thus

$$D(z) = \left(z^2 + \left(r + \frac{1}{r}\right)z + 1\right)K, \tag{2.37}$$

For $|a| < 2$ the unit circle has two complex conjugate zeros, and

$$D(z) = (z^2 + (2 \cos x)z + 1)K. \tag{2.38}$$

The special case for $a = 2$ is not of lowest order because it can be factored into the square of equation (2.35).

Any length-4 even-symmetric filter can be factored into products of terms of the form of (2.35) and (2.36).

The fourth case is an even-symmetric length-5 filter of the form

$$D(z) = z^4 + az^3 + bz^2 + az + 1.$$

For $a^2 < 4(b - 2)$ and $b > 2$, the zeros are neither real nor on the unit circle; therefore, they must have complex conjugates and must have images about the unit circle. The form of the transfer function is

$$D(z) = \left\{ z^4 + \left[\left(2\,\frac{r^2 + 1}{r} \right) \cos x \right] z^3 + \left[r^2 + \frac{1}{r^2} + 4 \cos^2 x \right] z^2 \right.$$

$$\left. + \left[\left(2\,\frac{r^2 + 1}{r} \right) \cos x \right] z + 1 \right\} K. \tag{2.39}$$

If one of the zeros of a length-5 filter is on the real axis or on the unit circle, $D(z)$ can be factored into a product of lower-order terms of the forms in (2.35), (2.37), and (2.38); therefore, it is not of lowest order.

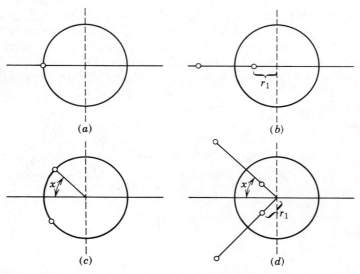

FIGURE 2.8. Zero locations for the basic linear-phase FIR filter transfer functions on the Z plane.

The odd-symmetric filters of (2.23) are described by the foregoing factors plus the basic length-2 filter described by

$$D(z) = (z - 1)K. \tag{2.40}$$

The zero locations for the four basic cases of type 1 and type 2 FIR filters are shown in Fig. 2.8. The locations for the type 3 and type 4 odd-symmetric cases of (2.23) are the same, plus the zero at unity from (2.40).

We can conclude from this analysis that all linear-phase FIR filters have zeros either on the unit circle or in the reciprocal symmetry of (2.37) or (2.39) about the unit circle and that their transfer functions can always be factored into products of terms with these four basic forms. This factored form can be used in implementing a filter by cascading short filters to realize a long filter. Knowing the locations of the zeros of the transfer function helps in developing programs for filter design and analysis.

SUMMARY

This chapter derived the basic characteristics of the FIR filter. For the linear-phase case the frequency response can be calculated easily. The effects of the linear phase can be separated so that the amplitude can be approximated as a real-valued function. This property is very useful for filter design. It was shown that there are four basic types of linear-phase FIR filters, each with character-istics important for design. The frequency response can be calculated by applying the DFT to the filter coefficients or, for greater resolution, to the N filter coefficients with zeros added to increase the length. An efficient calculation of the DFT uses the fast Fourier transform (FFT). The frequency response can also be calculated by special formulas that include the effects of linear phase.

Because of the linear-phase requirements, the zeros of the transfer function must lie on the unit circle in the z plane or they must occur in reciprocal pairs around the unit circle. This fact gives insight into the effects of the zero locations on the frequency response and can be used in the implementation of the filter.

The FIR filter is attractive from several viewpoints. It alone can achieve exactly linear phase. It is easily designed with linear methods. It cannot be unstable. The implementation or realization in hardware or on a computer is basically the calculation of an inner product, which can be accomplished efficiently. On the negative side, the FIR filter may require a rather long length to achieve certain frequency responses. Hence, a large number of arithmetic operations per output value and a large number of coefficients have to be stored. The linear-phase characteristic makes the time delay of the filter equal to half its length, which may be large.

How the FIR filter is implemented and whether it is chosen over alternatives depend strongly on the hardware or computer to be used. If an array processor is used, an FFT implementation[3] would probably be selected. If the TMS320 signal processor is used, a direct calculation of the inner product is probably best. If a VAX or similar general-purpose computer with floating-point arithmetic is used, an IIR filter may be chosen over the FIR, or the implementation of the FIR might take into account the symmetries of the filter coefficients to reduce arithmetic. To make these choices, one must consider the characteristics developed in this chapter together with the results developed later in this book.

REFERENCES

[1]   L. R. Rabiner and B. Gold, *Theory and Application of Digital Signal Processing*, Englewood Cliffs, NJ: Prentice-Hall, 1975.

[2]   A. V. Oppenheim and R. W. Schafer, *Digital Signal Processing*, Englewood Cliffs, NJ: Prentice-Hall, 1975.

[3]   C. S. Burrus and T. W. Parks, *DFT/FFT and Convolution Algorithms*, New York: Wiley-Interscience, 1985.

3

Design of Linear-Phase
Finite Impulse-Response

In this chapter, various methods of linear-phase FIR filter design are developed and analyzed.[1-5] The basic filter design problem involves the following steps:

1. Choose a desired ideal response, usually described in the frequency domain.
2. Choose an allowed class of filters (e.g., a length-N FIR filter).
3. Establish a measure or criterion of "goodness" for the response of an allowed filter compared to the desired response.
4. Develop a method to find the best member of the allowed class of linear-phase FIR filters as measured by the criterion of goodness.

This approach is often used iteratively. After the best filter is designed and evaluated, the desired response or the allowed class or the measure of quality might be changed; the filter would then be redesigned and reevaluated.

The Ideal Low-Pass Frequency Response
This chapter develops design procedures by considering the basic low-pass filter. The simplest ideal response has a pass band extending from $\omega = 0$ up to $\omega = \omega_1$ and a stop band extending from $\omega = \omega_1$ up to the Nyquist frequency of $\omega = \pi$ (see Fig. 3.1a). In some cases there is a region between the pass band and the stop band where neither the desired nor the undesired signals exist. Or the region may be the overlap of the spectra of the desired and undesired signals. This region can be defined as a transition region, with the ideal response having a transition function that more smoothly connects the pass-band and stop-band responses and allows a better total approximation (see Fig. 3.1b). The third possibility (Fig. 3.1c) does not define the approximation over the transition

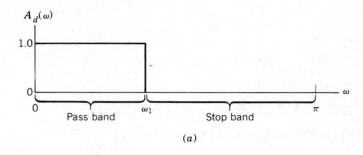

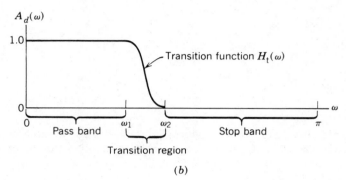

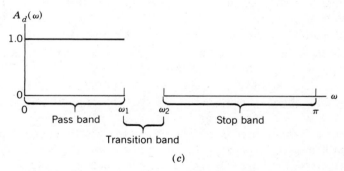

FIGURE 3.1. Ideal low-pass FIR filter frequency responses.

region, and it is called a transition band. All three cases are considered by the different design criteria and methods.

The Approximation Criteria

Three error measures are generally used in FIR filter design. One is the average of the squared error in the frequency-response approximation. The second is the maximum of the error over specified regions of the frequency response. A third approach is based on a Taylor series approximation to the desired response. The method based on the first error measure is called a least squared (LS) approximation, the second a Chebyshev approximation, and the third a

Butterworth or maximally flat approximation. Most of the useful design procedures are based on one of these three approximations, or on a combination of them, or on a modification of them.

The Design Methods

Each of the various methods for FIR filter design discussed in this chapter has some advantages and some disadvantages. The frequency-sampling method is fast and simple. It is useful for adaptive filters or for an intermediate stage in a more complicated algorithm where speed is important. Unfortunately, it gives the least control over the total frequency response.

The LS error methods use an error criterion that is related to the energy of the signal or noise, and the design equations are linear. However, the designs sometimes have frequency responses with oscillations or overshoots that may be undesirable. The use of windows is a simple method of controlling these effects, but it is rather ad hoc and the results are not optimal according to any known criterion. Use of a transition region or weights gives excellent results, but the problem may not have an analytical solution, or it may require the solution of ill-conditioned equations.

The Parks–McClellan algorithm minimizes the Chebyshev error, but the design algorithm can be slow. If smoothness of the response is needed, the maximally flat approximation has an analytical solution for the basic low-pass filter. A newly developed method based on Zolotarev polynomials gives a mixture of maximally flat and Chebyshev approximations. This chapter considers each of these methods and their characteristics.

3.1 FREQUENCY-SAMPLING DESIGN

The most straightforward design method is simply the inverse of the analysis procedure of (2.7) given in Section 2.1. The analysis calculates samples of the frequency response from the filter coefficients. This problem is well posed if N samples of a desired frequency response are used to find the N filter coefficients by simply solving the N simultaneous equations given by (2.8). This approach can also be viewed as an interpolation problem where the designed filter will have a frequency response that exactly passes through the desired points and, between those points, takes on values given by (2.5).

Directly solving the N equations of (2.8) is generally undesirable. Solving N simultaneous equations requires on the order of N^3 arithmetic operations, and the equations are sometimes ill conditioned. If the frequency samples are equally spaced, the DFT can be used. Since the DFT of the impulse response gives samples of the frequency response, the inverse DFT (IDFT) of samples of a desired frequency response will give the impulse response. This requires N^2 arithmetic operations in general, but only $N \log N$ operations if the FFT can be used.[13]

Since $h(n)$ is real and, for the linear-phase problem, symmetric, the required arithmetic is reduced by a factor of 4 compared to the direct DFT approach, and simple design formulas that have good numerical properties can be derived.

To develop explicit formulas for frequency-sampling design of linear-phase FIR filters, a direct use of the inverse DFT is most straightforward. For N equally spaced frequency-response samples of $C_k = H(2\pi k/N)$, the length-N FIR filter coefficients are given by the IDFT as

$$h(n) = \frac{1}{N} \sum_{k=0}^{N-1} C_k e^{j2\pi nk/N}. \tag{3.1}$$

When $H(\omega)$ is linear phase, (3.1) may be simplified by the formulas in Chapter 2 for the four types of linear-phase FIR filters. For example, the frequency response (2.26) for the type 1 filter for N is odd, $L = N$, and $M = (N - 1)/2$, and a frequency sample at $\omega = 0$ is

$$A_k = \sum_{n=0}^{M-1} 2h(n)\cos\left(\frac{2\pi(M - n)k}{N}\right) + h(M). \tag{3.2}$$

Using the amplitude function $A(\omega)$, defined in (2.10), of the form (3.2) and the IDFT (3.1) give for the impulse response

$$h(n) = \frac{1}{N} \sum_{k=0}^{N-1} e^{-j2\pi Mk/N} A_k e^{j2\pi nk/N}$$

$$= \frac{1}{N} \sum_{k=0}^{N-1} A_k e^{j2\pi(n - M)k/N}. \tag{3.3}$$

Because $h(n)$ is real, $A_k = A_{N-k}$ and (3.3) becomes

$$h(n) = \frac{1}{N}\left[A_0 + \sum_{k=1}^{M} 2A_k \cos\left(\frac{2\pi(n - M)k}{N}\right)\right]. \tag{3.4}$$

Only M of the $h(n)$ need be calculated because of the symmetries in (2.18).

This formula calculates the impulse response values $h(n)$ from the desired frequency samples A_k and requires M^2 operations rather than N^2. An interesting observation is that not only are (3.2) and (3.4) a pair of analysis and design formulas, but they are also a transform pair. Indeed, they are of the same form as a cosine transform.[14]

A similar development applied to the cases for even N from (2.27) gives the frequency samples

$$A_k = \sum_{n=0}^{N/2-1} 2h(n)\cos\left(\frac{2\pi(M - n)k}{N}\right). \tag{3.5}$$

The design formula becomes

$$h(n) = \frac{1}{N}\left[A_0 + \sum_{k=1}^{N/2-1} 2A_k \cos\left(\frac{2\pi(n-M)k}{N}\right)\right],$$ (3.6)

which is of the same form as (3.4), except that the upper limit on the summation recognizes N as even and (from Section 2.2.1) $A_{N/2} = 0$.

The scheme just described uses frequency samples at

$$\omega = \frac{2\pi k}{N}, \qquad k = 0, 1, 2, \ldots, N-1,$$ (3.7)

which are N equally spaced samples starting at $\omega = 0$. Another possible frequency-sampling scheme that allows design formulas has no sample at $\omega = 0$ but uses N equally spaced samples located at

$$\omega = \frac{(2k+1)\pi}{N}, \qquad k = 0, 1, 2, \ldots, N-1.$$ (3.8)

This form of frequency sampling is more difficult to relate to the DFT than the sampling of (3.5), but it can be done by stretching[15] A_k and taking a $2N$-length DFT.

The two cases for odd and even lengths and the two for samples at zero but not at zero frequency give a total of four cases for the frequency-sampling design method applied to linear-phase FIR filters of types 1 and 2, as defined in Section 2.2.1. For an odd length and no zero sample, we derive the analysis and design formulas analogously to (3.2) and (3.4):

$$A_k = \sum_{n=0}^{M-1} 2h(n)\cos\left(\frac{2\pi(M-n)(k+\frac{1}{2})}{N}\right) + h(M).$$ (3.9)

The design formula becomes

$$h(n) = \frac{1}{N}\left[\sum_{k=0}^{M-1} 2A_k\cos\left(\frac{2\pi(n-M)(k+\frac{1}{2})}{N}\right) + A_M \cos \pi(n-M)\right].$$ (3.10)

The fourth case, for an even length and no zero frequency sample, gives the analysis formula

$$A_k = \sum_{n=0}^{N/2-1} 2h(n)\cos\left(\frac{2\pi(M-n)(k+\frac{1}{2})}{N}\right),$$ (3.11)

and the design formula

$$h(n) = \frac{1}{N}\left[\sum_{k=0}^{N/2-1} 2A_k \cos\left(\frac{2\pi(n-M)(k+\frac{1}{2})}{N}\right)\right].$$ (3.12)

Formulas (3.4), (3.6), (3.10), and (3.12) allow a straightforward design of the four frequency-sampling cases. They and their analysis companions in (3.1), (3.5), (3.9), and (3.11) also are the four forms of discrete cosine and inverse cosine transforms. A FORTRAN program that implements these four designs is given as Program 1 in the appendix.

The designs of even-symmetric linear-phase FIR filters of types 1 and 2 in Section 2.2.1 have been developed here. A similar development for the odd-symmetric filters, types 3 and 4, can easily be performed, with the results closely related to the discrete sine transform. Using the frequency sampling scheme of (3.7), we obtain the type 3 analysis and design results:

$$A_k = \sum_{n=0}^{M-1} 2h(n)\sin\left(\frac{2\pi(M-n)k}{N}\right),$$

$$h(n) = \frac{1}{N}\left[\sum_{k=1}^{M} 2A_k \sin\left(\frac{2\pi(M-n)k}{N}\right)\right]. \tag{3.13a}$$

For type 4

$$A_k = \sum_{n=0}^{N/2-1} 2h(n)\sin\left(\frac{2\pi(M-n)k}{N}\right),$$

$$h(n) = \frac{1}{N}\left[\sum_{k=1}^{N/2-1} 2A_k \sin\left(\frac{2\pi(M-n)k}{N}\right) + A_{N/2}\sin(\pi(M-n))\right]. \tag{3.13b}$$

If we use the frequency-sampling scheme of (3.8), the type 3 equations become

$$A_k = \sum_{n=0}^{M-1} 2h(n) \sin\left(\frac{2\pi(M-n)(k+\frac{1}{2})}{N}\right),$$

$$h(n) = \frac{1}{N}\left[\sum_{k=0}^{M-1} 2A_k \sin\left(\frac{2\pi(M-n)(k+\frac{1}{2})}{N}\right)\right]. \tag{3.13c}$$

For type 4

$$A_k = \sum_{n=0}^{N/2-1} 2h(n)\sin\left(\frac{2\pi(M-n)(k+\frac{1}{2})}{N}\right)$$

$$h(n) = \frac{1}{N}\left[\sum_{k=0}^{N/2-1} 2A_k \sin\left(\frac{2\pi(M-n)(k+\frac{1}{2})}{N}\right)\right]. \tag{3.13d}$$

These type 3 and type 4 formulas are useful in the design of differentiators and Hilbert transformers[1,2,9,31] directly and as the base of the discrete LS error methods in Section 3.2.1.

3.1.1 Guidelines for Frequency-Sampling Design

Guidelines are necessary for choosing among the four cases for frequency-sampling design. Some examples may aid in making the choice. One example is a low-pass filter with a pass band extending through half of the range to the Nyquist frequency or folding frequency (one half of the sampling frequency). With normalized notation, the sampling frequency is $\omega = 2\pi$ rad/s or $f = 1$ Hz (one cycle per second). As shown in Table 2.1, the frequency response is periodic with period 2π if N is odd and period 4π if N is even. Therefore, the maximum frequency or Nyquist frequency is $\omega = \pi$ or $f = 0.5$ Hz.

A linear-phase FIR filter is designed to approximate the ideal low-pass response that has a pass band from $\omega = 0$ to $\pi/2$ and a stop band from $\omega = \pi/2$ to π. The ideal amplitude frequency-response plot is shown in Fig. 3.1a.

Experience shows that the total frequency response of the designed filter becomes closer to the ideal as the length becomes longer. However, the measure of closeness must be carefully defined. For now, we assume that an approximate length has been chosen. The choice of whether the length is odd or even is made by matching the intrinsic properties of the response of an even or odd length, as shown in Table 2.1 and Fig. 2.3, to the desired response. Finally, the choice of a frequency sample at zero frequency or not is made. This choice is made to make the transition between pass band and stop band fall as near as possible to halfway between two sample points. The design can then be calculated by one of the formulas in (3.4), (3.6), (3.10), or (3.12).

If frequency-sampling design with an ideal desired frequency response having a discontinuity causes too much oscillation or overshoot between the samples, a transition region can be added to the ideal response. That is discussed in Section 3.2.3.1. The shape of the transition function can have an important influence on the design.

Example 3.1. Length-21 Low-pass Filter by Frequency Sampling

This example concerns the design of an odd length-21, linear-phase, low-pass filter where the desired frequency response has a pass band that is half of the maximum frequency, as illustrated in Fig. 3.1a. Thus, the band edge is $f_p = 0.25$ Hz for a sampling frequency of one sample per second. The odd-length formula (3.4) for a sample at zero frequency is used to design the filter (see Program 1 in the appendix). The resulting filter coefficients are

$$h(0) = h(20) = -0.0324800,$$
$$h(1) = h(19) = 0.0381875,$$
$$h(2) = h(18) = 0.0288168,$$
$$h(3) = h(17) = -0.0476191,$$
$$h(4) = h(16) = -0.0264265,$$
$$h(5) = h(15) = 0.0651707,$$
$$h(6) = h(14) = 0.0249165,$$

$$h(7) = h(13) = -0.1070000,$$
$$h(8) = h(12) = -0.0240785,$$
$$h(9) = h(11) = 0.3186069,$$
$$h(10) = 0.5238096,$$

The frequency response is shown in Fig. 3.2a and 3.2b. Note the ripples in the pass band and the stop band near the band edge and the exact interpolation of the sample points in the pass band and the stop band. The maximum stop-band attenuation is approximately 16 db. The locations of the zeros of the transfer function are shown in Fig. 3.2c. A total of 20 zeros for the twentieth-degree polynomial is formed from the 21 filter coefficients in (2.4). There are 10 zeros on the unit circle that come from the samples in the stop band and have the form of (2.38), as shown in Fig. 2.8c. There are 10 zeros with 8 occuring in two sets of four, as given in (2.39) and shown in Fig. 2.8d and two on the real axis, as in Fig. 2.8b and given in (2.37).

Example 3.2. Length-20 Low-pass Filter by Frequency Sampling

This example uses the same specifications as Example 3.1 with a band edge of $f_p = 0.25$ Hz, but uses an even-length design of $N = 20$ from (3.5). The filter

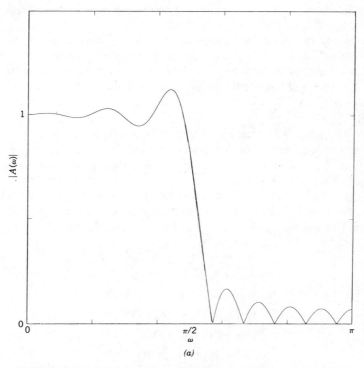

FIGURE 3.2. Length-21 low-pass FIR filter by frequency sampling.

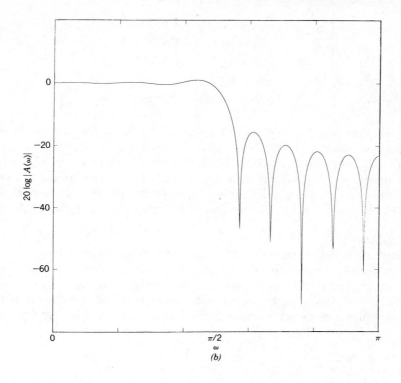

(b)

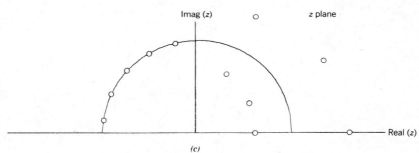

(c)

FIGURE 3.2. (Continued)

coefficients from (3.6) and Program 1 in the appendix are

$$h(0) = h(19) = -0.03257,$$
$$h(1) = h(18) = 0.04384,$$
$$h(2) = h(17) = 0.02071,$$
$$h(3) = h(16) = -0.05702,$$
$$h(4) = h(15) = -0.00516,$$

$$h(5) = h(14) = \quad 0.07675,$$
$$h(6) = h(13) = -0.02234,$$
$$h(7) = h(12) = -0.12071,$$
$$h(8) = h(11) = \quad 0.11191,$$
$$h(9) = h(10) = \quad 0.48459,$$

The frequency response is given in Fig. 3.3a and 3.3b, and the transfer function zeros are shown in Fig. 3.3c. Note the zero at $\omega = \pi$ that all even-length filters have, in contrast to the case for $N = 21$. This results in a single zero on the unit circle on the real axis since there is now a total of nine zeros on the unit circle. The pass-band and stop-band performances of this example are very close to those in Example 3.1, but the location of the band edge is slightly higher.

Extensions
A possible generalization with the frequency-sampling design is the specification of the desired frequency-response samples over a nonuniform spacing of frequencies. The IDFT cannot be used, and specific design formulas cannot be derived for this case, but the design problem can be posed by taking N samples of the frequency responses given in (2.19), (2.21), or (2.24) at arbitrary frequencies

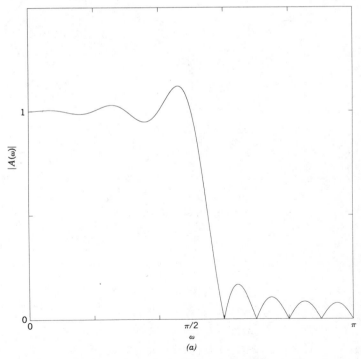

FIGURE 3.3. Length-20 low-pass FIR filter by frequency sampling.

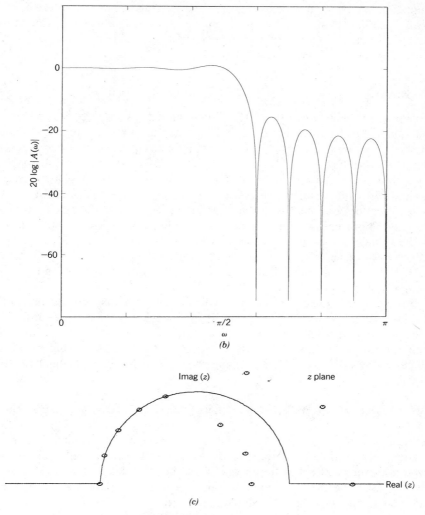

FIGURE 3.3. (Continued)

to give N simultaneous equations that can be solved for $h(n)$. More generally, for arbitrary-phase response, (2.5) can be sampled at N frequencies. Again, care must be used to avoid large errors between sample frequencies.

The frequency-sampling design method can be used directly to design FIR filters; however, it may also be used as a starting point or intermediate stage in a more complicated method. It is also used to design the high-order prototype filters that are truncated in the LS error method of Section 3.2.1.

Section 3.2.2.1 shows that by allowing a transition region as in Fig. 3.1b, it is possible to reduce the overshoot or ripple in the pass band and stop band for a given order. It is also possible to further reduce the ripple by al'owing several sample points in the transition region and adjusting their value to minimize an error criterion.[1] This was a popular approach to FIR filter design before the Remes exchange algorithm was developed for the direct design of optimal Chebyshev filters.

Summary

This section developed the FIR filter frequency-sampling design method specifically for the linear-phase filter. Direct design with the frequency-sampling method is possible by applying the inverse DFT to equally spaced samples of the frequency response $H(\omega)$. This direct use of the IDFT can design arbitrary-phase as well as linear-phase FIR filters. If it is used to design linear-phase filters, the desired $H(\omega)$ must satisfy the constraints given in Section 2.2 and the linear phase must be consistent with the filter length. Great care must be exercised in designing both linear- and nonlinear-phase filters by the direct method. If the desired phase response is in some way inconsistent with the magnitude or the filter length, large errors will occur between the sample points. However, when used with care and when checked by analysis, it is a very simple and powerful design tool.

For the linear-phase FIR filter, simple design formulas were developed that automatically take care of the phase. These formulas are easy to use and can be implemented on small computers or calculators to design very long filters. The design formulas or the IDFT give very little numerical error. Using the FFT to calculate the IDFT gives a fast design procedure with even less numerical error.

3.2 LEAST SQUARED ERROR FREQUENCY-DOMAIN DESIGN

The purpose of most filters is to separate desired signals from undesired signals or noise. Often the descriptions of the signals and noise are given in terms of their frequency content or the energy of the signals in frequency bands. For this reason, filter specifications are generally given in the frequency domain, and, since the energy of a signal is related to the square of the signal, a squared error approximation criterion is often appropriate. This section considers two methods of defining a squared error. The first definition is the sum of the squares of the error measured at a finite set of frequency sample points. The second is the integral of the square of the error over a finite or infinite range of frequencies.

3.2.1 Discrete Frequency Samples

The frequency-sampling design method is really not an approximation approach but an interpolation method that produces a filter with a frequency response that exactly passes through the sample points. However, there is no constraint on the response between the sample points, and poor results may be obtained. In this section we control the response between sample points by considering a number of sample points larger than the order of the filter. Because this results in more equations than unknowns, only approximate solutions are possible.

The frequency response of an FIR filter presented in (2.5) is given by

$$H(\omega) = \sum_{n=0}^{N-1} h(n)e^{-j\omega n}.$$

The design problem is posed by defining an error measure E as a sum of the squared differences between the actual and the desired frequency response over a set of L frequency samples. This error function is defined as

$$E = \sum_{k=0}^{L-1} |H(\omega_k) - H_d(\omega_k)|^2, \tag{3.14}$$

where $H_d(\omega_k)$ are the L samples of the desired response. This problem is easier to formulate and solve if the frequency samples are equally spaced, which gives

$$\omega_k = \frac{2\pi k}{L},$$

and the problem is restricted to linear-phase filters, where the real-valued amplitude $A(\omega)$, rather than the complex frequency response $H(\omega)$, can be approximated. For approximations to a complex response, see Chapter 4.

With these conditions (3.14) becomes

$$E = \sum_{n=-(L-1)/2}^{(L-1)/2} |h_n - h_{dn}|^2,$$

or, with a simpler notation,

$$E = \sum_{k=0}^{L-1} |A_k - A_{dk}|^2. \tag{3.15}$$

A powerful property of the Fourier transform permits a straightforward design of LS error FIR filters. Parseval's theorem,[13-16] based on the orthogonality of the DFT, states that the error defined by (3.15) in the frequency domain can also be calculated in the time domain for L odd by

$$E = \sum_{n=-(L-1)/2}^{(L-1)/2} |h_n - h_{dn}|^2,$$

where h_{dn} is the length-L FIR filter that has the L frequency-response amplitude samples A_{dk}. We can calculate this filter by the frequency-sampling method of

Section 3.1, using the special formulas such as (3.4) for length L or the IDFT. A factor of $1/L$ is omitted from these equations to simplify the development. The filter to be designed has a length-N impulse response h_n with L frequency-response samples A_k. Because the filter is of length-N, the error equation can be split into two sums:

$$E = \sum_{n=-M}^{M} |h_n - h_{dn}|^2 + \sum_{n=M+1}^{(L-1)/2} 2|h_{dn}|^2.$$

This equation clearly shows that to minimize E, we need to choose the N values of h_n to be equal to the equivalent N values of h_{dn}. In other words, we obtain h_n by simply truncating h_{dn}. The second summation then gives the residual error. Examining the residual error as a function of N may help to choose the filter length N.

For the type 1 linear-phase FIR filter (described in Section 2.2) with odd length N and even-symmetric impulse response, the L equally spaced samples of the frequency response from (2.19) give (2.26) and (3.1). The samples are

$$A_k = \sum_{n=0}^{M-1} 2h(n)\cos\left(\frac{2\pi(M-n)k}{L}\right) + h(M)$$

$$\text{for } k = 0, 1, 2, \ldots, L-1,$$

where $M = (N-1)/2$. This formula was derived as a special case of the DFT applied to the type 1, real, even-symmetric FIR filter coefficients to calculate the sampled amplitude of the frequency response. We noted in Section 3.1 that it is also a cosine transform, and it can be shown that this transformation is orthogonal over the independent values of A_k, just as the DFT is.

To use the alternative equally spaced sampling in (3.8), which has no sample at zero frequency, we must calculate h_{dn} from (3.10). The type 2 filters with even N are developed in a similar way and use the design formulas (3.6) and (3.12). These methods are summarized as follows:

The filter design procedure for an odd-length filter is to first design an odd-length-L FIR filter by the frequency-sampling method from (3.4) or (3.10) or the IDFT, then to symmetrically truncate it to the desired odd-length N. To design an even-length filter, start with an even-length-L frequency-sampling design from (3.6) or (3.12) or the IDFT and symmetrically truncate. The resulting length-N FIR filters are an optimal LS error approximation to the desired frequency response over the L samples.

This approach can also be applied to the general arbitrary-phase FIR filter design problem discussed in Chapter 4.

It is sometimes desirable to formulate the mean-error design problem using unequally spaced frequency samples and/or a weighting function on the error. This formulation requires a different approach to the solution.

Equation (2.19) relates the L frequency samples to the $M + 1$ independent values of the symmetric length-N impulse response $h(n)$. The design problem that gives the A_k and the values for $h(n)$ represents L equations with $M + 1$ unknowns. Because of the symmetries of $A(\omega)$ shown in Fig. 2.3, only half of the L values of A_k are independent; however, to have proper weights on all L samples, we must calculate all values.

Equation (2.19) sampled at L arbitrary frequencies can be written as a matrix equation

$$F\mathbf{h} = \mathbf{a}, \tag{3.16}$$

where $\mathbf{h}$ is an $M + 1$ length vector with elements that are the first half of $h(n)$. F is an L-by-$(M + 1)$ matrix of the cosine terms from (2.19), and $\mathbf{a}$ is a length-L vector of the frequency samples A_k. If the formula for the calculation of L values of the frequency response of a length-N FIR filter in (2.19) is used to define an error vector of differences, as defined in (3.15), and the result is written in the matrix formulation of (3.16), the error becomes

$$F\mathbf{h} = \mathbf{a} = \mathbf{a}_d + \mathbf{e} \tag{3.17}$$

or

$$F\mathbf{h} - \mathbf{a}_d = \mathbf{e},$$

where $\mathbf{e}$ is a vector of differences between the actual and desired samples of the frequency response. The error measure defined in (3.15) becomes the quadratic form

$$E = \mathbf{e}^T\mathbf{e}. \tag{3.18}$$

For $L > N$, equation (3.16) is overdetermined and cannot, in general, be solved for h. The filter design error measure is the norm of $\mathbf{e}$, as given in (3.18). This error measure is minimized by making $\mathbf{e}$ orthogonal to the columns of F in (3.17). Multiplying both sides of (3.17) by the transpose of F gives

$$F^TF\mathbf{h} = F^T\mathbf{a}_d + F^T\mathbf{e}.$$

For E to be minimum, $\mathbf{e}$ must be orthogonal to the columns of F and, therefore, $F^T\mathbf{e}$ must be zero. The optimal $\mathbf{h}$ must satisfy the "normal equations"[7]

$$F^TF\mathbf{h} = F^T\mathbf{a}_d. \tag{3.19}$$

Equation (3.19) can be rewritten in terms of the pseudoinverse[7,8] as

$$\mathbf{h} = [F^TF]^{-1}F^T\mathbf{a}_d. \tag{3.20}$$

If $L = N$, (3.20) becomes the regular frequency-sampling problem and can be solved with zero error. For the case of interest in this section, where $L > N$, there are still only $M + 1$ equations to be solved. For $L \gg N$, equation (3.19) may be ill conditioned, and (3.20) should not be used to solve it. Special methods will be necessary to avoid serious numerical problems.[8]

If a weighted error function is desired, (3.15) is modified to give

$$E = \sum_{k=0}^{L-1} W_k |A(\omega_k) - A_d(\omega_k)|^2. \tag{3.21}$$

The normal equations of (3.19) become

$$F^T W F \mathbf{h} = F^T W \mathbf{a}_d, \tag{3.22}$$

where W is a positive-definite matrix of the weights. If zero weights are desired, the effect is achieved by removing those frequencies from the set of L frequencies, not by using a zero-value weight, which would violate the vector space conditions of a well-posed minimization problem.

Although developed here for the linear-phase filter, (3.22) is a general design approach for the FIR filter that allows arbitrary-phase sampling as well as uneven frequency sampling and a weighting function in the error definition. For the arbitrary-phase case a complex F is obtained from sampling (2.5). For the special case of the equally spaced frequency samples and linear-phase filter with unity weighting, the solution of (3.19) or (3.22) is the same as given by the frequency-sampling design formulas in (3.4).

An important use of the unequally spaced frequency samples is the creation of a transition band between the pass band and the stop band where there are no samples. This "don't care" band does not contribute to the error measure E and allows much better approximation to occur over the pass band and stop band.

Of the many ways to solve (3.19) or (3.22), one of the easiest and most reliable is the linear algebra software package LINPACK,[8] which has a special program to solve this least mean squared error problem. Equation (3.20) should not be solved directly. For large L it is ill conditioned, and a direct solution will probably have large errors. LINPACK uses special algorithms to minimize these numerical errors.

This approach was applied to the same problems that were solved by frequency sampling in the previous section. For $N = L$ the same results were obtained, thus verifying the theoretical prediction. As L becomes larger compared to N, more control is exerted over the behavior between the original sample points, and the solution approaches the same results as obtained in the next section, where the error is defined as a continuous function of frequency and the integral of the squared error is minimized. A program that calls LINPACK to design a linear-phase FIR filter by these methods is Program 2 in the appendix. Although the solution of the normal equations is a powerful and flexible technique, it can be slow, have numerical problems, and require large amounts of computer memory.

Example 3.3. Low-pass Filter Designed by Discrete Least Squared Error

The same desired frequency response as used in Example 3.1 with a band edge of $f_p = 0.25$ Hz is used with the discrete LS error design method of (3.19). Program 2 in the appendix is used to find the length-21 filter coefficients. Over 81 frequency samples are optimized. The coefficients can also be found from (3.4) and Program 1; use a length of 81 and truncate to length 21. The coefficients are

$$
\begin{aligned}
h(0) = h(20) &= -0.00629, \\
h(1) = h(19) &= 0.03555, \\
h(2) = h(18) &= 0.00625, \\
h(3) = h(17) &= -0.04561, \\
h(4) = h(16) &= -0.00621, \\
h(5) = h(15) &= 0.06376, \\
h(6) = h(14) &= 0.00619, \\
h(7) = h(13) &= -0.10616, \\
h(8) = h(12) &= -0.00618, \\
h(9) = h(11) &= 0.31833, \\
h(10) &= 0.50617.
\end{aligned}
$$

Figures 3.4a and 3.4b illustrate the frequency response and Fig. 3.4c the zero locations for the length-21 filter. The filter has slightly less pass-band ripple and a minimum stop-band attenuation of 20 db. The zeros in the stop band are no longer equally spaced as they were for the frequency-sampling design. The simple frequency-sampling design forces the zeros and ripples to be equally spaced. By this not being the case for the LS error design, it obtains less pass-band ripple and more stop-band attenuation simultaneously. Note the zero locations compared to those in Fig. 3.2c.

Example 3.4. Low-pass Filter with a Transition Region Designed by Discrete Least Squared Error

Figure 3.5 illustrates the frequency response of a length-21 filter designed with a transition region and a linear transition function. The pass band goes from $f = 0$ to $f = 0.2$, the transition region is from $f = 0.2$ to $f = 0.3$ with a linear transition function, and the stop band is from $f = 0.3$ to the Nyquist frequency $f = 0.5$ Hz. Note the reduction in the overshoot near the band edges as compared with Example 3.3. We used Program 2 to design this example, optimizing over 199 samples, with a modified section for the desired response. The same result could also be obtained by truncating a frequency-sampling design. The filter coefficients are

$$
\begin{aligned}
h(0) = h(20) &= -0.00004, \\
h(1) = h(19) &= 0.00362, \\
h(2) = h(18) &= -0.00172, \\
h(3) = h(17) &= -0.01635, \\
h(4) = h(16) &= 0.00377,
\end{aligned}
$$

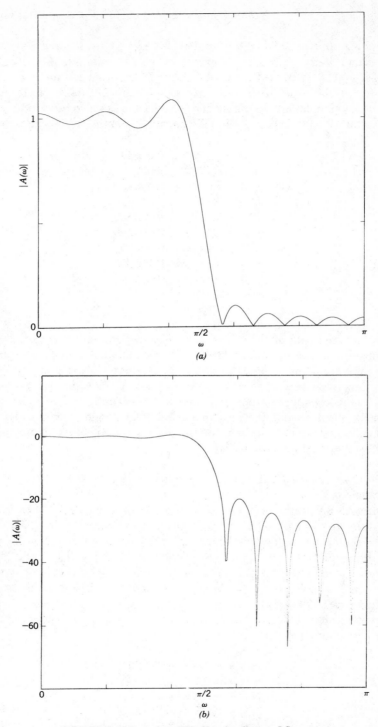

FIGURE 3.4. Low-pass FIR filter by discrete LS error.

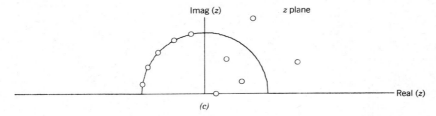

Imag (z)   z plane

Real (z)

(c)

FIGURE 3.4. (Continued)

$$h(5) = h(15) = \quad 0.04013,$$
$$h(6) = h(14) = -0.00569,$$
$$h(7) = h(13) = -0.09077,$$
$$h(8) = h(12) = \quad 0.00705,$$
$$h(9) = h(11) = \quad 0.31299,$$
$$h(10) = \quad 0.49246.$$

Example 3.5. Low-pass Filter with a Transition Band Designed by Discrete Least Squared Error over the Pass Band and Stop Band

Figure 3.6 illustrates the frequency response of a length-21 filter designed with unequally spaced frequency samples so that the transition region is not included in the error at all. There are 336 frequency samples equally spaced over the same pass band and stop band as in Example 3.4. Note the further reduction in overshoot, which is the result of putting no constraints on the transition region response. The filter coefficients were calculated from Program 9 in the appendix, with the ideas from (3.19), and are

$$h(0) \; = h(20) = \quad 0.00009,$$
$$h(1) \; = h(19) = \quad 0.00871,$$
$$h(2) \; = h(18) = -0.00021,$$
$$h(3) \; = h(17) = -0.02120,$$
$$h(4) \; = h(16) = \quad 0.00032,$$
$$h(5) \; = h(15) = \quad 0.04418,$$
$$h(6) \; = h(14) = -0.00044,$$
$$h(7) \; = h(13) = -0.09342,$$
$$h(8) \; = h(12) = \quad 0.00051,$$
$$h(9) \; = h(11) = \quad 0.31394,$$
$$h(10) = \quad 0.49945.$$

Summary

This section formulated an FIR filter design problem based on an LS error criterion and developed two methods of solution. The first method requires the samples of the desired frequency response to be equally spaced and the error to

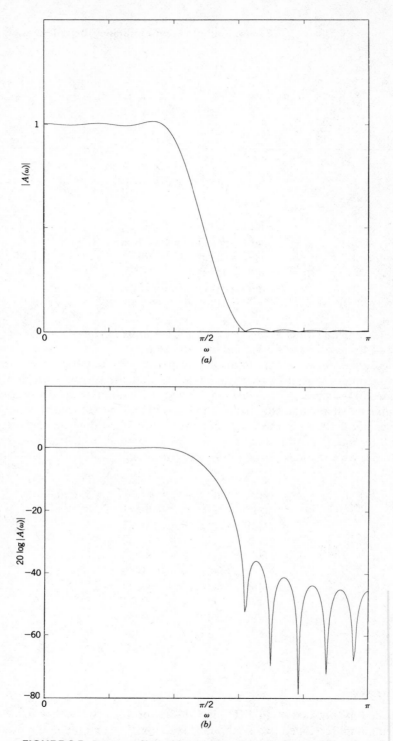

FIGURE 3.5. Low-pass filter with transition region by discrete LS error.

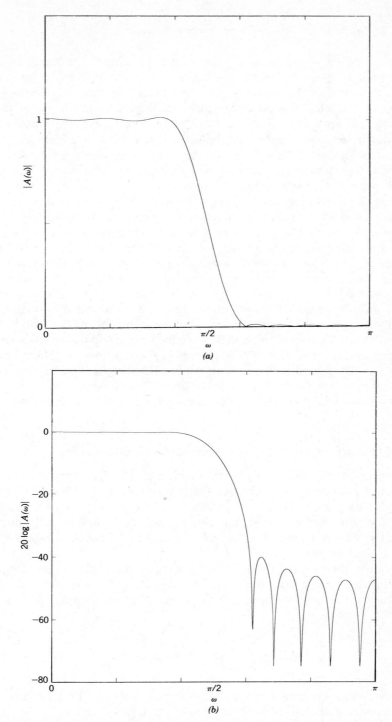

FIGURE 3.6. Low-pass filter with a transition band by discrete LS error over the pass band and stop band.

have no weighting function. When these conditions are met, a length-L IDFT can be used to design an FIR filter, which is truncated to length N to give an LS error FIR filter from an arbitrary, desired frequency response. If a linear-phase FIR filter is desired, special design formulas for the length-L prototype are available and are implemented in Program 1. In both cases the problems are numerically well conditioned, and the design calculations are fairly fast. If the FFT can be used for the IDFT, they can be very fast. Because this method can be applied to an arbitrary desired frequency response, the excessive oscillation that occurs near a discontinuity in the desired response, known as the Gibbs phenomenon, can be reduced by using a transition region between the pass band and stop band with a transition function to remove the discontinuity.

The second method of solution requires solving the normal equations, which are a set of overdetermined simultaneous equations. This formulation is more general in that it allows unequally spaced frequency samples and an error weight function, but it is slower and often numerically ill conditioned. The solution is best achieved by using special algorithms that minimize the inherent numerical errors of this approach. The FORTRAN program in the appendix (Program 2) solves this problem with the software subroutines in LINPACK.[8] This method allows the use of a transition band between the pass band and stop band, which does not contribute at all to the error measure. It also allows weighting the error to get a better approximation in some regions. The length of a filter that can be designed by this method is limited by the size of the computer memory available and by numerical errors. When it can be used, the results are excellent.

3.2.2 Integral Squared Error Approximation Criterion

In the previous section the value of the amplitude of the frequency response was controlled by using an error function that was defined over L frequency samples where L was greater than the filter length N. In certain cases, such as for the basic low-pass filter, it is possible to find an analytical solution to the problem where the error is defined as an integral over all of the frequency response. This section develops that case.

An error measure is defined as the integral of the square of the difference between the actual amplitude and the desired amplitude over the basic frequency range of $-\pi < \omega < \pi$. This measure is called the *integral squared error*[1-4].

$$E = \frac{1}{2\pi} \int_{-\pi}^{\pi} |A(\omega) - A_d(\omega)|^2 \, d\omega. \qquad (3.23)$$

The result of minimizing this error measure will be close to that obtained from minimizing the discrete error measure in (3.15) for very large L.

If we assume that the filter impulse response $h(n)$ is infinite in duration and symmetric about the origin, then there is zero phase shift, and the amplitude of the frequency response can be written as

$$H(\omega) = A(\omega) = \sum_{n=-\infty}^{\infty} h(n)e^{-j\omega n}, \tag{3.24}$$

which is simply the Fourier transform of $h(n)$, with an inverse of

$$h(n) = \frac{1}{2\pi} \int_{-\pi}^{\pi} A(\omega)e^{j\omega n} \, d\omega.$$

Parseval's theorem[13-16] states that the energy in a signal can be calculated in the time domain as well as in the frequency domain. Equation (3.23) becomes

$$E = \frac{1}{2\pi} \int_{-\pi}^{\pi} |A(\omega) - A_d(\omega)|^2 \, d\omega,$$

$$= \sum_{n=-\infty}^{\infty} |h_n - h_{dn}|^2,$$

where h_{dn} is the inverse transform of $A_d(\omega)$, calculated by (3.25), and h_n is the length-N filter with frequency response $A(\omega)$. Because h_n is of finite length N, the error summation can be split into two sums

$$E = \sum_{n=-M}^{M} |h_n - h_{dn}|^2 + \sum_{n=M+1}^{\infty} 2h_{nd}^2,$$

where $M = (N-1)/2$. This expression clearly shows that to minimize E, the N values of h_n should be chosen equal to the corresponding N values of h_{dn}. The residual error for the optimal h_n is given by the second summation.

Another interpretation of these equations is that (3.24) is a Fourier series expansion of the periodic function $A(\omega)$, and the second equation is the formula for the series coefficients. From the theory of Fourier series we know that a truncated series is an optimal approximation to the expanded function in the sense that the integral squared error is minimized.

For the type 1 linear-phase FIR filter described in Section 2.2, which has odd length N and an even-symmetric impulse response, the amplitude frequency response from (3.24) is

$$A(\omega) = h_0 + \sum_{n=1}^{\infty} 2h_n \cos(\omega n),$$

with an infinite, noncausal impulse response of

$$h(n) = \frac{1}{\pi} \int_{0}^{\pi} A(\omega)\cos(\omega n) \, d\omega. \tag{3.25}$$

The type 2 filter, which has an even length, can be designed by modifying this approach, which is illustrated in the development for the ideal low-pass filter later in this section. The design method is summarized as follows:

If the impulse response of a linear-phase FIR filter is found by symmetrically truncating the Fourier series expansion of the desired amplitude response, the resulting filter will have an amplitude response that is an LS error approximation to the desired response.

This method is similar to the results obtained in the last section with the discrete LS error criterion and can be used either analytically or numerically for any desired amplitude response that satisfies the conditions required of a linear-phase FIR filter as given in Section 2.2.1.

The main limitation of this design method is the difficulty in calculating h_{dn} from $A_d(\omega)$. This problem exists because the calculation requires evaluating the integral in (3.25) rather than the sum in (3.4) for the discrete frequency formulation of the LS error design. For only a few practical desired frequency responses can a formula be derived. Fortunately, it is possible for the basic low-pass filter with desired amplitude given in Figs. 3.1a and 3.1b.

For the low-pass filter we assume that the desired amplitude is unity from $\omega = 0$ to $\omega = W\pi$, and zero from $\omega = W\pi$ to $\omega = \pi$, as shown in Fig. 3.7.

For N odd the base frequency range for the coefficient equation (3.25) is $-\pi$ to π (or 0 to 2π). The desired amplitude is 1 from $-W\pi$ to $W\pi$, where W is the cutoff frequency of the filter expressed as a fraction of the total range from 0 to π (see Fig. 3.7). The ideal or desired amplitude is given over the frequency range $-\pi < \omega < \pi$ by

$$A(\omega) = \begin{cases} 1 & \text{for } -W\pi < \omega < W\pi, \\ 0 & \text{otherwise.} \end{cases} \tag{3.26}$$

From (3.25) the impulse response

$$h(n) = \frac{1}{\pi} \int_0^{W\pi} \cos(\omega n)\, d\omega$$

becomes

$$h(n) = \frac{\sin(W\pi n)}{\pi n}, \tag{3.27}$$

FIGURE 3.7. Ideal low-pass filter response for odd N.

which has even symmetry about the origin and therefore is noncausal. Since it is infinitely long, it is not physically realizable. However, this $h(n)$ gives the exact desired frequency response of (3.26) and Fig. 3.7. To make this result physically possible, we truncate the impulse response of (3.27) (symmetrically to maintain the linear-phase property), which, according to Fourier series theory, gives a finite-length impulse response with an amplitude that is an LS error approximation to the desired response. The finite-length $h(n)$ is shifted to the right to make it causal. The shift only adds a linear phase and does not destroy the minimum error property.

To obtain a length-N filter, we truncate the impulse response by setting all terms for $n < -M$ and $n > M$ equal to zero, where $M = (N-1)/2$, as defined in (2.16). The response is then shifted to the right by M terms, and the result is a causal, optimal LS error FIR filter. The resulting impulse response is

$$h(n) = \begin{cases} \dfrac{\sin(W\pi(n-M))}{\pi(n-M)}, & 0 \leqslant n \leqslant N-1, \\ 0 & \text{otherwise.} \end{cases} \tag{3.28}$$

If the transition between the pass band and the stop band is expressed as f_1 in hertz rather than W as a fraction of the $0 < \omega < \pi$ region, and if the sampling rate is one sample per second, (3.28) becomes

$$h(n) = \begin{cases} \dfrac{\sin(2\pi f_1(n-M))}{\pi(n-M)}, & 0 \leqslant n \leqslant N-1, \\ 0 & \text{otherwise.} \end{cases} \tag{3.29}$$

See Fig. 3.8.

The derivation of $h(n)$ for even N is slightly more complicated because there is no zero-phase, even-length FIR filter corresponding to (3.27) and Fig. 3.8a. The amplitude response of the ideal low-pass filter with even N is given in Fig. 3.9. By using a frequency range of -2π to 2π, we obtain a double-length $h(n)$ with zero values at the even indices. It is compressed to length N by simply removing the zero values and then truncated and shifted to give an even-length optimal filter. The process is illustrated in Fig. 3.10.

The design formula for the even-length case is exactly the same as for the odd length given in (3.28), but note that M is now a fraction.

Example 3.6. Length-21 Low-pass Filter Designed by Least Squared Error
This example is the straightforward, continuous, LS error design given by (3.28) with simple truncation. The frequency response and zero locations given in Fig. 3.11 are very similar to those resulting from the discrete LS error method in Fig. 3.4 for a band edge of $f_p = 0.25\,\text{Hz}$. The filter coefficients calculated by

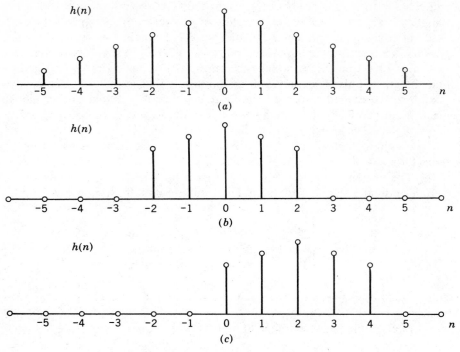

FIGURE 3.8. Design of optimal LS error FIR filter.

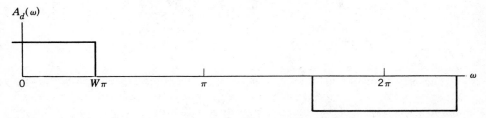

FIGURE 3.9. Ideal low-pass filter response for even N.

Program 4 in the appendix with a rectangular window are

$$
\begin{aligned}
h(0) &= h(20) = &&0.00000, \\
h(1) &= h(19) = &&0.03537, \\
h(2) &= h(18) = &&0.00000, \\
h(3) &= h(17) = &&-0.04547, \\
h(4) &= h(16) = &&0.00000, \\
h(5) &= h(15) = &&0.06366, \\
h(6) &= h(14) = &&0.00000, \\
h(7) &= h(13) = &&-0.10610, \\
h(8) &= h(12) = &&0.00000, \\
h(9) &= h(11) = &&0.31831, \\
& h(10) = &&0.50000.
\end{aligned}
$$

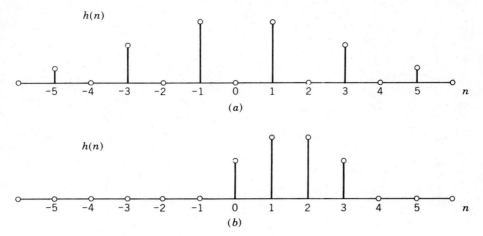

FIGURE 3.10. Design of optimal LS error FIR filter for even N.

Note that the even-indexed terms are all zero except for the center term $h(10)$. This situation occurs because the band edge is exactly one half of the Nyquist frequency. This advantage is important in implementing the filter because these multiplications need not be carried out. In practical applications a possible change in the band edge or the sampling rate should always be considered as a method to reduce arithmetic. This design is the classic LS error low-pass FIR filter and it should be compared to the other FIR filter designs in this book.

The Gibbs Phenomenon
As shown in this and previous examples, the amplitude frequency response of the low-pass filter has an oscillating behavior that is more pronounced near the edge of the pass band. This behavior is known as the *Gibbs phenomenon*[15,16] and is the result of approximating a discontinuity in the desired frequency response. Early in the study of Fourier series, it was found that if a function with a discontinuity was approximated by a Fourier series, there would be an overshoot in the region near the discontinuity. As the number of Fourier series terms increased, the squared error decreased and approached zero as the number of terms approached infinity. However, the maximum value of the overshoot, and therefore the maximum value of the error, did not go to zero but approached a constant value of approximately 11% of the size of the discontinuity. See Fig. 3.12. This behavior is exactly what happens in the case of the LS error design of a low-pass FIR filter. Although it is less well known, it also happens in the frequency-sampling design method where it approaches approximately 18% of the discontinuity. See Fig. 3.13.

The Gibbs phenomenon overshoot may be undesirable, but it is a direct consequence of minimizing the squared error when approximating a discontinuity with no transition region. Any reduction of the overshoot increases the value of the squared error. This basic conflict of desired properties causes a rethinking of the whole formulation of the LS error design problem.

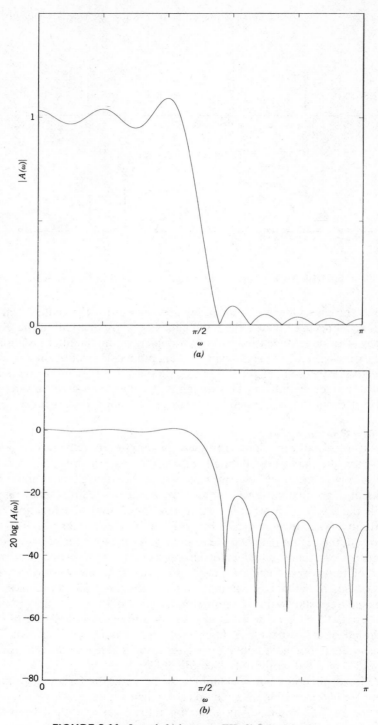

FIGURE 3.11. Length-21 low-pass FIR filter by LS error.

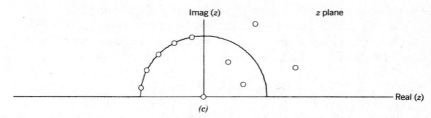

(c)

FIGURE 3.11. (Continued)

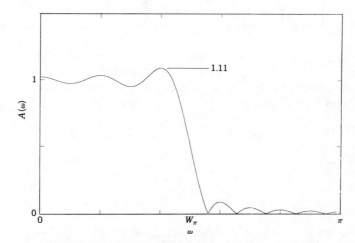

FIGURE 3.12. The Gibbs phenomenon in LS error filter design.

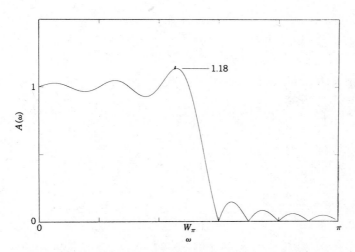

FIGURE 3.13. Example of overshoot for frequency-sampling design.

61

Least Squared Error Approximation of a Differentiator

The preceding development has considered the approximation of an ideal low-pass filter. The ideal differentiator can also be well approximated by a linear-phase FIR filter. The frequency response of a differentiator is

$$H(\omega) = j\omega \quad \text{or} \quad A(\omega) = \omega.$$

Because $A(\omega)$ is an odd function of ω and there is a constant $90°$ phase shift, the design should use type 3 or type 4, as defined in Section 2.2.1 and described in Table 2.1. If the differentiation is to be combined with a low-pass filter to reduce high-frequency interference, type 3 should be used because it always has a zero response at $A(\pi)$. If the widest possible bandwidth is desired, type 4 should be used.

For the design of a type 3 differentiator that has odd length N, $A_d(\omega)$ is defined over $-\pi < \omega < \pi$ and is periodic with period 2π, which was the case for the low-pass filter in (3.26) shown in Fig. 3.7. The filter coefficients are derived from (3.1) and are given by

$$h(n) = -\frac{1}{2\pi} \int_{-\pi}^{\pi} \omega \sin(\omega n)\, d\omega$$

$$= \begin{cases} \dfrac{\cos(\pi n)}{n}, & n \neq 0, \\ 0 & n = 0. \end{cases} \tag{3.30}$$

For the design of a type 4 wide-band differentiator with even N, we define $A_d(\omega)$ over $-2\pi < \omega < 2\pi$, which is periodic with period 4π, as was the case for the low-pass filter of Fig. 3.9. The ideal response is

$$A_d(\omega) = \begin{cases} -2\pi - \omega, & -2\pi < \omega < -\pi, \\ \omega & -\pi < \omega < \pi, \\ 2\pi - \omega, & \pi < \omega < 2\pi. \end{cases}$$

The filter coefficients are

$$h(n) = -\frac{1}{4\pi} \int_{-2\pi}^{2\pi} A_d(\omega)\sin\left(\frac{\omega n}{2}\right) d\omega,$$

$$h(n) = -\frac{1}{\pi} \int_{0}^{\pi} \omega \sin\left(\frac{\omega n}{2}\right) d\omega, \quad n = \pm 1, \pm 3, \pm 5, \ldots, \pm \infty,$$

$$h(n) = -\frac{4 \sin(\pi(2n + 1)/2)}{\pi(4n^2 + 4n + 1)}, \quad n = 0, \pm 1, \pm 2, \ldots. \tag{3.31}$$

These results must be truncated and shifted, as for the low-pass filter, to give a length-N causal filter. An implementation of these two differentiator designs is given in the appendix in Program 5. The Gibbs phenomenon occurring in the type 3 approximation can be reduced by truncating with windows (see Section 3.2.3.4), but the LS error optimality will be lost.

Examples 3.7 and 3.8. A Least Squared Error Design of Type 3 and Type 4 Differentiators
     Figure 3.14 shows the frequency response of a type 3 differentiator designed by Program 5 from (3.30) and truncated for length 21. Note the zero response at $f = 0.5$, which is characteristic of all type 3 filters. Figure 3.15 shows the frequency response of a length-20, type 4 differentiator from (3.31). Note the wider frequency range of the approximation and the reduction in overshoot even though the length is shorter than the type 3 example. The filter coefficients for the differentiators are

Length-21 differentiator	Length-20 differentiator
$-h(0)\ \ = h(20) = -0.10000$	$-h(0) = h(19) = -0.00353$
$-h(1)\ \ = h(19) = \ \ \ 0.11111$	$-h(1) = h(18) = \ \ \ 0.00441$
$-h(2)\ \ = h(18) = -0.12500$	$-h(2) = h(17) = -0.00566$
$-h(3)\ \ = h(17) = \ \ \ 0.14286$	$-h(3) = h(16) = \ \ \ 0.00753$
$-h(4)\ \ = h(16) = -0.16667$	$-h(4) = h(15) = -0.01052$
$-h(5)\ \ = h(15) = \ \ \ 0.20000$	$-h(5) = h(14) = \ \ \ 0.01572$
$-h(6)\ \ = h(14) = -0.25000$	$-h(6) = h(13) = -0.02598$
$-h(7)\ \ = h(13) = \ \ \ 0.33333$	$-h(7) = h(12) = \ \ \ 0.05093$
$-h(8)\ \ = h(12) = -0.50000$	$-h(8) = h(11) = -0.14147$
$-h(9)\ \ = h(11) = \ \ \ 1.00000$	$-h(9) = h(10) = \ \ \ 1.27324$
$-h(10)\ \ \ \ \ \ \ \ \ \ \ \ = 0.00000$	

Least Squared Error Approximation with a Transition Region
The FIR filter design problem can be made much more versatile and flexible by introducing a transition region between the pass band and the stopband, as illustrated in Fig. 3.1b. This formulation fits the way filter specifications are usually given much better than using one frequency to specify the separation of the pass band and stop band. Also, the Gibbs phenomenon can be eliminated, and the approximation in the pass band and stop band can be improved by the transition region.
     If a transition function is defined as part of the ideal response to connect the unity pass-band response and zero stop-band response, the shape of this function can be chosen to minimize the approximation error for a given length. The pass band is defined as $0 < f < f_1$, the transition region as $f_1 < f < f_2$, and the stop band as $f_2 < f < 0.5$. The transition function $H_t(f)$ is defined over the transition region. A development similar to that for the simple no-transition-region case can give analytical formulas for optimal LS error $h(n)$ for several interesting transition functions. This ideal response is shown in Fig. 3.16, with the frequency f given in hertz.

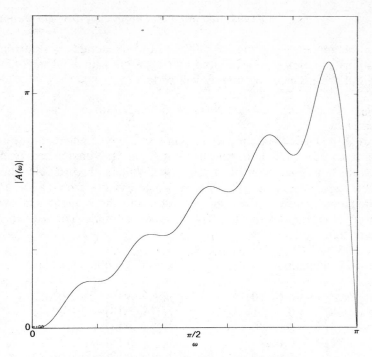

FIGURE 3.14. Length-21 FIR differentiator by LS error.

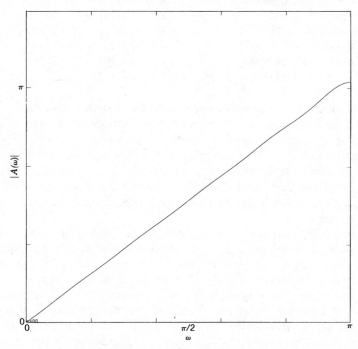

FIGURE 3.15. Length-20 FIR differentiator by LS error.

If the transition function is a simple straight line (first-order spline) connecting the pass-band response to the stop-band response, the impulse response is

$$h(n) = \begin{cases} \dfrac{\sin(\pi(f_2 - f_1)(n - M))}{\pi(f_2 - f_1)(n - M)} \dfrac{\sin(\pi(f_2 + f_1)(n - M))}{\pi(n - M)}, & 0 \leqslant n \leqslant N - 1, \\ 0, & \text{otherwise.} \end{cases}$$

(3.32a)

We can derive (3.32a) directly from the inverse Fourier transform of the desired ideal response $A_d(\omega)$ by using (3.25), in much the same way as (3.28) was developed. An alternative approach is to observe that the ideal response of Fig. 3.16, which has a first-order spline (straight-line) transition function, can be created by convolving an ideal rectangular response whose band edge is at $(f_2 + f_1)/2$ with a narrow rectangular function whose width is $f_2 - f_1$ and whose height is $1/(f_2 - f_1)$. This approach gives the inverse Fourier transform of the final desired function as being 2π times the product of the inverse transforms of the two rectangles. That can easily be seen by comparing the second term in (3.23a) to (3.29) with a band edge of the average of f_1 and f_2 and noting that the first term is the same as (3.29) with a total width of $f_2 - f_1$.

If the transition function is a second-order spline (two sections of parabolas), the impulse response is

$$h(n) =$$

$$\begin{cases} \left(\dfrac{\sin(\pi(f_2 - f_1)(n - M)/2)}{\pi(f_2 - f_1)(n - M)/2}\right)^2 \dfrac{\sin(\pi(f_2 + f_1)(n - M))}{\pi(n - M)}, & 0 \leqslant n \leqslant N - 1, \\ 0 & \text{otherwise.} \end{cases}$$

(3.32b)

We can also derive (3.32b) directly from (3.25) or indirectly by convolving two half-width rectangles to obtain a triangle function of width $f_2 - f_1$, which is then convolved with the basic ideal filter response to get the desired transition function.

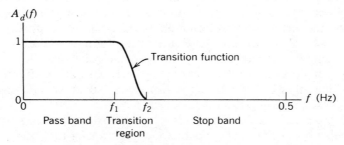

FIGURE 3.16. Ideal response with a transition region.

We can easily generalize (3.32b) to higher-order spline transition functions. An optimal LS error approximation to an ideal response with a P-order spline transition function is given by

$$h(n) =$$

$$\begin{cases} \left(\dfrac{\sin(\pi(f_2 - f_1)(n - M)/P)}{\pi(f_2 - f_1)(n - M)/P}\right)^P \dfrac{\sin(\pi(f_2 + f_1)(n - M))}{\pi(n - M)}, & 0 \leqslant n \leqslant N - 1, \\ 0 & \text{otherwise.} \end{cases}$$

(3.32c)

Equation (3.32c) is derived by convolving together P rectangles of width $1/P$ and then convolving the result with the basic filter response. It is possible to create other P-order splines by using unequal widths as long as the final width gives the correct transition width.

An alternative transition function that can result in an analytical solution uses sections of trigonometric functions. One useful function is the raised cosine defined by

$$A_d(f) = \begin{cases} 1, & 0 < f < f_1, \\ 1 + \cos\left(\dfrac{\pi(f - f_1)}{(f_2 - f_1)}\right), & f_1 < f < f_2, \\ 0, & f_2 < f < 0.5. \end{cases}$$

This definition gives an FIR filter with coefficients given by

$$h(n) = \begin{cases} \dfrac{\cos(\pi(f_2 - f_1)(n - M)}{1 - 4(f_2 - f_1)^2(n - M)^2} \dfrac{\sin(\pi(f_2 + f_1)(n - M))}{\pi(n - M)} & 0 < n < N \\ 0 & \text{otherwise} \end{cases}$$

(3.33)

This function can be generalized by adding higher-frequency cosine terms to give a smoother transition. It can also be combined with the spline functions to give a very rich class of possible transition functions for flexible design.

The faster the coefficients $h(n)$ decrease with increasing n, the smaller the error that will result when the inverse Fourier transform is truncated. Fourier theory[15,16] shows that the smoother $A(\omega)$ is, the faster $h(n)$ drops off with increasing n. If $A(\omega)$ can be differentiated Q times with finite results, $h(n)$ will drop off as a multiple of $1/n$ to the $(Q + 1)$st power. Note that is the case for the results of (3.28) and (3.30)–(3.33).

Using a transition region with an LS error approximation design procedure gives a much more flexible and useful method, yet it retains the optimality of the designed filter. It is, however, ad hoc in the sense that the transition function must be chosen by experience and trial. The choice of transition function depends on the transition width, the bandwidth, and the filter length N. The result is in the form of a weighted version of the simple no-transition design, which is similar to the result of using windows (discussed in Section 3.2.3.4), but it is more directly related to the specifications and optimality criterion.

The concept of transition regions can also be used to design other than low-pass filters. For example, it can be used with the differentiator, Hilbert transform, high-pass, and other designs.

FORTRAN Program 3 designs least integral squared error linear-phase FIR filters with an ideal low-pass response and a transition region. It allows a choice of P-order splines and a raised cosine transition function.

Example 3.9. Least Squared Error Design of a Low-pass FIR Filter with First-Order Spline Transition Function

Figure 3.17 shows the frequency response of a filter with $f_p = 0.2$ and $f_s = 0.3$ Hz, the same pass-band and stop-band specifications as Examples 3.4 and 3.5. This filter was designed from (3.32a). A first-order spline transition function and Program 3 were used. The filter coefficients are

$$
\begin{aligned}
h(0) &= h(20) = & 0.00000, \\
h(1) &= h(19) = & 0.00387, \\
h(2) &= h(18) = & 0.00000, \\
h(3) &= h(17) = & -0.01673, \\
h(4) &= h(16) = & 0.00000, \\
h(5) &= h(15) = & 0.04053, \\
h(6) &= h(14) = & 0.00000, \\
h(7) &= h(13) = & -0.09108, \\
h(8) &= h(12) = & 0.00000, \\
h(9) &= h(11) = & 0.31310, \\
h(10) & = & 0.50000.
\end{aligned}
$$

Summary

This section defined an integral squared error and described a design procedure. One reason why the mean squared error criterion is useful is that it is a measure of energy. The power of a signal is a function of the square of the signal. That is easily seen when the signal is a voltage or current, or perhaps a force or velocity. Since the energy of a signal is the integral of its power, the integral squared error is proportional to the energy of the error. Other considerations are sometimes important, such as the maximum value of the error. Unfortunately, filters designed to minimize the integral squared error often do not have good maximum error characteristics if the desired response has discontinuities or rapid changes and no transition region.

The basic ideal low-pass filter design problem can be solved analytically and

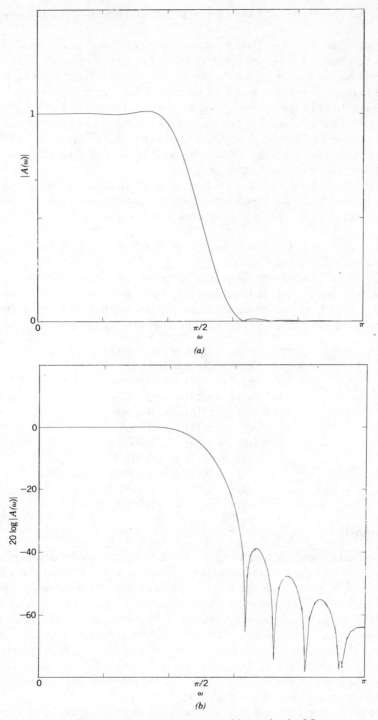

FIGURE 3.17. Low-pass filter with a transition region by LS error.

a formula can be derived. This analytical result is simple yet powerful. It allows us to design arbitrary even- or odd-length, optimal LS error, linear-phase FIR filters by a formula easily evaluated on a pocket calculator. A length-21 example is given as Example 3.4.

Analytical design formulas were derived for LS error approximations to ideal differentiators. Two cases were presented: one for odd length, where there is necessarily a zero response at $\omega = \pi$, and one for the wide-band case. These results could be extended to combine with the low-pass filter or to incorporate transitions regions or other modifications.

Introducing a transition region and a transition function into the formulation of the ideal frequency response produced considerable improvement in the approximation and added flexibility in specifying the filter. Analytical solutions were developed for several interesting cases with spline and trigonometric transition functions. This method is a powerful design algorithm. Program 3, which designs LS error low-pass filters with a transition region, is given in the appendix.

3.2.3 Transition Regions, Weighting Functions, and Windows for FIR Filter Design

Four approaches can improve the characteristics of filters designed by minimizing the squared error and can reduce the overshoot occurring near a discontinuity. The most straightforward solution is simply to change the desired frequency response so that there is no discontinuity and, therefore, no Gibbs phenomenon. This method has already been introduced and is easily carried out by having a transition region in the frequency response between the pass-band region and the stop-band region. That would allow a transition function for the desired response that would connect the pass-band and stop-band ideal responses.

The second approach is to change the error criterion in such a way as to reduce or remove the overshoot. That can be done by removing a region from the optimization. That region is then called a transition band or "don't care" region. We can do it by using unequally spaced samples in the discrete LS error method.

The third approach changes the error measure by introducing a weighting function in (3.23) to weight the error more where there is overshoot and/or less over regions that are not as important, such as the transition region.

The fourth method uses the result of a regular LS error design, such as (3.14), (3.28), (3.30), or (3.32), and directly modifies h_n to reduce the overshoot, but the result will no longer be optimal. Since the overshoot is caused by truncating the finite length-L sequence (3.14) or the infinite sequence (3.25), it can be reduced by a more gentle truncation achieved with time-domain windows.[1-4]

These ideas, which look at design in a broader sense, can be applied not only to continuous LS error problems, defined in Section 3.2.1, but to the discrete LS error problem of Section 3.2.2.

3.2.3.1 Modification of the Desired Frequency Response

In most filtering applications there is a range of frequencies that contain the spectrum of the desired signal, called the *pass band*, and a range that contains the undesired signals or noise, called the *stop band*. To ease the design problem, we define a transition band between the pass band and the stop band. The wider we make this band, the better we can make the approximations in the pass band and stop bands. The simplest modification to the desired amplitude response is to connect the unity gain in the pass band to the zero gain in the stop band by a straight line (see Fig. 3.1*b*). The desired amplitude response is

$$A_d(\omega) = \begin{cases} 1 & \text{pass band} & (0 < \omega < \omega_1), \\ \dfrac{\omega_2 - \omega}{\omega_2 - \omega_1} & \text{transition region} & (\omega_1 < \omega < \omega_2), \\ 0 & \text{stop band} & (\omega_2 < \omega < \pi). \end{cases}$$

The solution to the integral LS error approximation problem was given in (3.30a). More complicated spline and trigonometric transition functions can give further improvement and are given in (3.30). With the discrete LS error method and the frequency sampling method, the introduction of the transition region can also significantly reduce the Gibbs phenomenon and give greater control over the design process.

3.2.3.2 Use of a Transition Band

One of the most effective modifications of the direct LS error design methods is to change the bands of frequencies over which the minimization is carried out. In Section 3.2.3.1 a transition function is defined over the transition region of frequencies to create a continuous ideal frequency response to be approximated. In this section the band of frequencies for the transition region is simply removed from the error definition, and the region is called the transition band or "don't care" band. The error in (3.23) becomes

$$E = \int_0^{\omega_1} |A(\omega) - A_d(\omega)|^2 \, d\omega + \int_{\omega_2}^{\pi} |A(\omega) - A_d(\omega)|^2 \, d\omega. \qquad (3.34)$$

This transition band will give less squared error and can give a greater reduction of the overshoot than a transition function can because there is no constraint placed on $A(\omega)$ in the transition region. Some sets of specifications, however, will result in strange behavior in the transition band. If "don't care" is stated, it must be intended. Some experimentation will probably be necessary. This kind of design is best performed interactively where the results of various modifications and weighting functions can be compared.

3.2.3.3 Use of a Weighted Mean Squared Error Criterion

Flexibility is added to the definition of error given in (3.23) by introducing a positive weighting function $W(\omega)$ to give

$$E = \frac{1}{\pi} \int_0^{\pi} W(\omega)|A(\omega) - A_d(\omega)|^2 \, d\omega,$$

similar to that used with the discrete LS error case in (3.21). With this new function, more weight can be given to the approximating error in regions of greater interest or importance. Using the transition band introduced in Section 3.2.3.2 can sometimes result in undesirable behavior of the designed filter in that band. In those cases it may be preferable to use a transition function with a small weight to place some control over the transition region. In some cases a transition band with weighting functions has advantages. As we noted before, it is very important to try and compare different design philosophies by some kind of interactive design and analysis system.

3.2.3.4 Use of Window Functions in the Design of FIR Filters

The truncation of the infinite or length-L time-domain sequence causes the Gibbs phenomenon at a discontinuity in the frequency domain. This truncation can be viewed as multiplying the prototype time-domain sequence by a rectangular function that has value unity for $-M < n < M$ and zero outside that range. If $h_d(n)$ is the ideal prototype sequence symmetric about 0 and $h(n)$ the finite truncated result, the process can be described by

$$h(n) = h_d(n)r(n), \tag{3.35}$$

where

$$r(n) = \begin{cases} 1, & |n| \leqslant M, \\ 0, & |n| > M, \end{cases} \tag{3.36}$$

with $M = (N - 1)/2$ for N odd. The infinite-duration impulse response $h_d(n)$, with values given by (3.27) and illustrated in Fig. 3.8a, is multiplied by $r(n)$ to give $h(n)$, as shown in Fig. 3.8b. The response $h_d(n)$ has the ideal desired frequency response of (3.26) and Fig. 3.1, and $h(n)$ has the realizable frequency response with the undesirable overshoot.

Multiplication of two functions in the time domain corresponds to convolution of the Fourier transforms of the two signals. Indeed, it is a good way to see what causes the Gibbs phenomenon. The Fourier transform of the rectangular function used for truncation is

$$R(\omega) = \text{FT}\{r(n)\} = \sum_{n=-\infty}^{\infty} r(n) \, e^{-j\omega n} \tag{3.37}$$

$$= \sum_{n=-M}^{M} e^{-j\omega n} = \frac{\sin(\omega N/2)}{\sin(\omega/2)}. \tag{3.38}$$

If the Fourier transforms of $h_d(n)$ and $h(n)$ are $H_d(\omega)$ and $H(\omega)$, respectively, the time-domain truncation operation given in (3.37) is described in the frequency

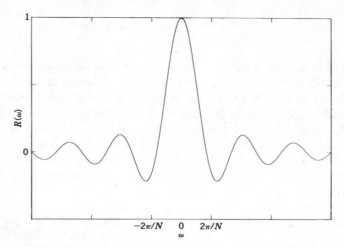

FIGURE 3.18. The Fourier transform of a rectangular function.

domain by

$$H(\omega) = H_d(\omega) * R(\omega), \tag{3.39}$$

which states that the frequency response of the finite-length filter is equal to the ideal frequency response convolved with a transform of the rectangular function given in (3.38). The frequency response of the width-N rectangular function $R(\omega)$ near $\omega = 0$ appears as shown in Fig. 3.18.

The result of convolving this oscillating function with the ideal low-pass frequency response gives the Gibbs phenomenon overshoot. We can see from Fig. 3.18 that as N gets larger, the width of the main part of the oscillating function grows narrower, although the height remains the same.

It is the existence of and size of the oscillating side "lobes" of $R(\omega)$ that cause the Gibbs phenomenon. An ideal $R(\omega)$ would be smooth and, therefore, cause no overshoot. It is the width of the main part of $R(\omega)$ that causes the slow transition from the pass band to the stop band. An ideal $R(\omega)$ would have zero width. This ideal function would be a Dirac delta function that, when convolved with the desired low-pass filter response, would introduce no change. For a finite-length filter this is impossible, but it does give an ideal for $R(\omega)$ to approximate.

Six standard windows found in the literature are[1-4,9,10]

1. *Bartlett triangular window:*

$$W(n) = \begin{cases} \dfrac{2(n+1)}{N+1}, & n = 0, 1, 2, \ldots, \dfrac{N-1}{2}, \\[3mm] 2 - \dfrac{2(n+1)}{N+1} & n = \dfrac{N-1}{2}, \ldots, N-1, \\[3mm] 0, & \text{otherwise.} \end{cases} \tag{3.40}$$

2–5. *Generalized cosine windows*

(rectangular, Hanning, Hamming, and Blackman):

$$W(n) = \begin{cases} a - b\cos\left(\dfrac{2\pi(n+1)}{N+1}\right) + c\cos\left(\dfrac{4\pi(n+1)}{N+1}\right), & n = 0, 1, 2, \ldots N-1 \\ 0, & \text{otherwise} \end{cases} \quad (3.41)$$

6. *Kaiser window with parameter* β:

$$W(n) = \begin{cases} \dfrac{I_0(\beta\sqrt{1 - (2(n+1)/(N+1))^2})}{I_0(\beta)}, & n = 0, 1, 2, \ldots, N-1, \\ 0, & \text{otherwise.} \end{cases} \quad (3.42)$$

The generalized cosine window has four special forms that are commonly used. These are determined by the parameters a, b, and c.

Window	a	b	c
Rectangular	1	0	0
Hanning	0.5	0.5	0
Hamming	0.54	0.46	0
Blackman	0.42	0.5	0.08

The most straightforward of these windows is the simple rectangular window, which gives the simple truncation and the classical Gibbs phenomenon. The Bartlett or triangular window reduces the overshoot but spreads the transition region considerably. The Hanning, Hamming, and Blackman windows use progressively more complicated cosine functions to provide a smooth truncation of the ideal impulse response and a frequency response that looks progressively better. The best window results probably come from using the Kaiser window, which has a parameter β that allows adjustment of the compromise between the overshoot reduction and transition region width spreading. Pass-band and stop-band ripple and transition width can be converted into a filter length N and a parameter β.[1] The Kaiser window requires calculating a Bessel function. A simple subroutine that evaluates this Bessel function is used in Program 4 in the appendix.

Plots of these window functions are shown in Fig. 3.19 to illustrate the various shapes that try to reduce the effects of the truncation without changing the basic characteristics of the LS error filter.[1-4]

Example 3.10. Length-21 Low-pass Filter with a Hanning Window

The filter designed in Example 3.6 is truncated by using a Hanning window described in (3.41) to give the results in Fig. 3.20. Note the smoother pass-band and greater stop-band attenuation but the wider transition region from pass

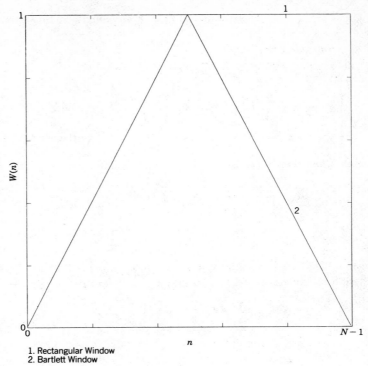

1. Rectangular Window
2. Bartlett Window

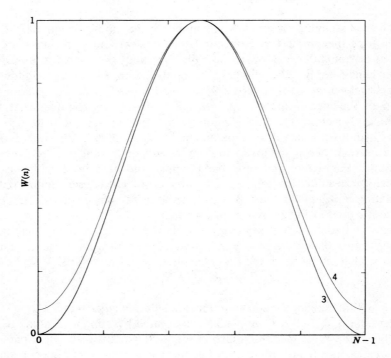

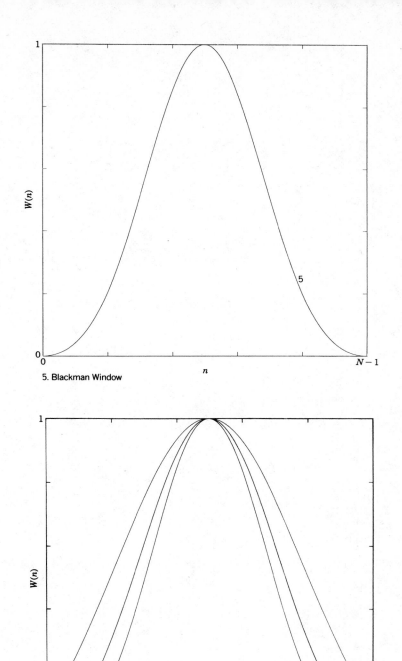

5. Blackman Window

FIGURE 3.19. Window functions for linear-phase FIR filter design.

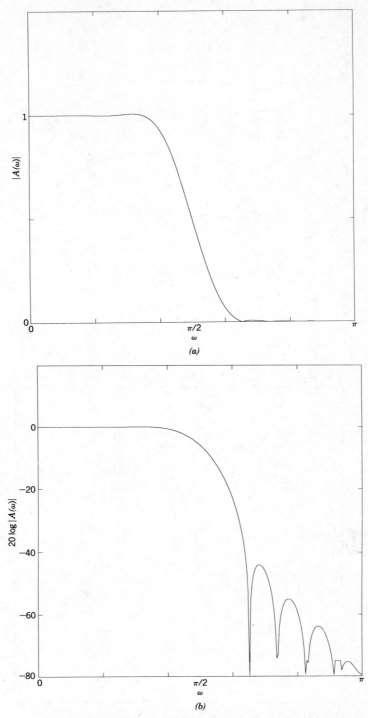

FIGURE 3.20. Length-21 low-pass filter with a Hanning window.

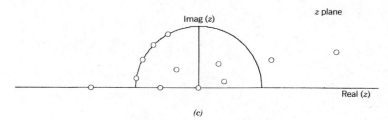

(c)

FIGURE 3.20. (Continued)

band to stop band. The filter coefficients are

$$
\begin{aligned}
h(0) &= h(20) = & 0.00000 \\
h(1) &= h(19) = & 0.00281 \\
h(2) &= h(18) = & 0.00000 \\
h(3) &= h(17) = & 0.01329 \\
h(4) &= h(16) = & 0.00000 \\
h(5) &= h(15) = & 0.03636 \\
h(6) &= h(14) = & 0.00000 \\
h(7) &= h(13) = & -0.08779 \\
h(8) &= h(12) = & 0.00000 \\
h(9) &= h(11) = & 0.31186 \\
h(10) &= & 0.50000
\end{aligned}
$$

Example 3.11. Length-21 Low-pass Filters with Kaiser Windows
The flexible Kaiser window in (3.42) is used with the parameter β equal to 4, 6.5, and 9 to give the results in Fig. 3.21, 3.22, and 3.23, respectively. Here, the pass band is very smooth, and the parameter β allows a tradeoff between transition width and stop-band attenuation.[1,4] The coefficients are

	$\beta = 4$	$\beta = 6.5$	$\beta = 9$
$h(0) = h(20) =$	0.00000	0.00000	0.00000
$h(1) = h(19) =$	0.00600	0.00142	0.00034
$h(2) = h(18) =$	0.00000	0.00000	0.00000
$h(3) = h(17) =$	-0.01750	-0.00848	-0.00413
$h(4) = h(16) =$	0.00000	0.00000	0.00000
$h(5) = h(15) =$	0.04033	0.02874	0.02054
$h(6) = h(14) =$	0.00000	0.00000	0.00000
$h(7) = h(13) =$	-0.09056	-0.08062	-0.07182
$h(8) = h(12) =$	0.00000	0.00000	0.00000
$h(9) = h(11) =$	0.31285	0.30892	0.30506
$h(10) =$	0.50000	0.50000	0.50000

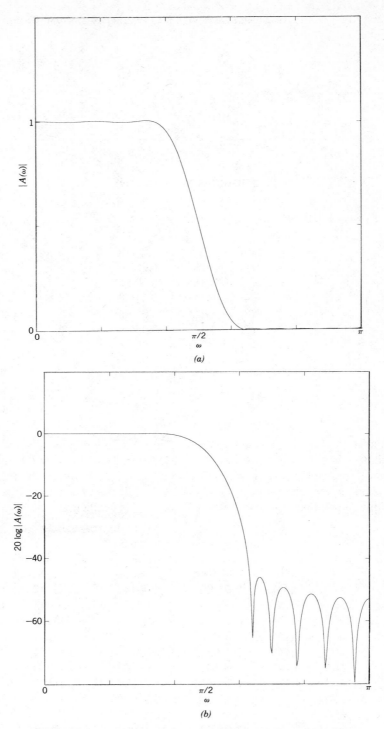

FIGURE 3.21. Length-21 low-pass filter with a Kaiser window, $\beta = 4$.

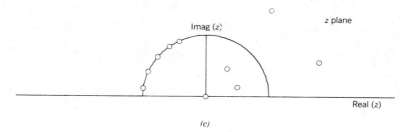

(c)

FIGURE 3.21. (Continued)

Compare the coefficients resulting from the various windows and note how fast they decrease as *n* increases.

Example 3.12. Length-101 Low-pass Filter with a Kaiser Window

To illustrate the effects of filter length, we designed a length-101 low-pass filter, using the same method as Example 3.6 but with a greater length. Compare the results in Fig. 3.21 to that for the length-21 filter in Fig. 3.24. Increasing the length improves all characteristics except, of course, the implementation problems.

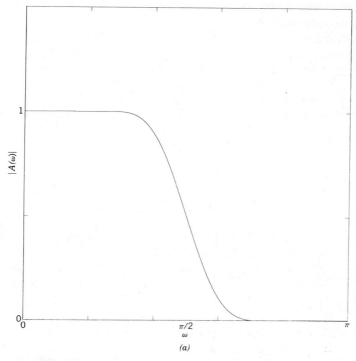

(a)

FIGURE 3.22. Length-21 low-pass filter with a Kaiser window, $\beta = 6.5$.

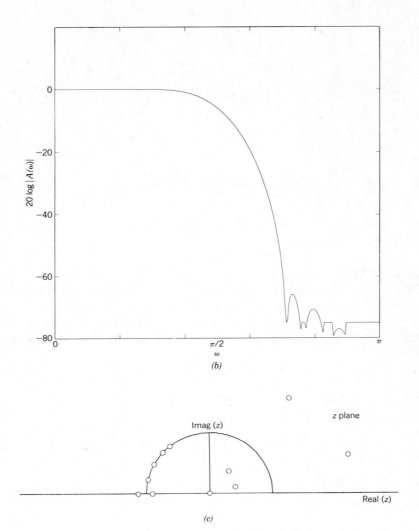

FIGURE 3.22. (Continued)

Summary

This section considered four methods of modifying the straightforward LS error FIR filter design. The simplest and probably the best is to change the ideal frequency response being approximated. Introducing a transition region with a transition function significantly reduces the resulting approximation ripple or Gibbs phenomenon and still allows the use of the special formulas and IDFT methods of Sections 3.1, 3.2.1, and 3.2.2.

Introducing a transition band with no contribution to the error measure requires solving the normal equations and does not allow using formulas or the IDFT because of unequally spaced frequency samples. This approach is more

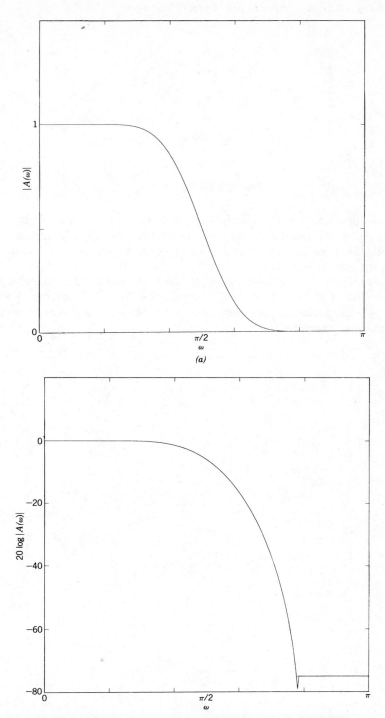

FIGURE 3.23. Length-21 low-pass filter with a Kaiser window, $\beta = 9$.

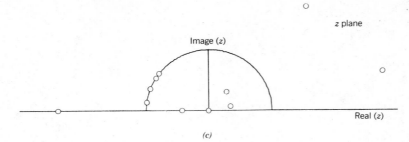

(c)

FIGURE 3.23. (Continued)

prone to numerical errors, is fairly slow, and requires considerable computer memory. However, when it can be used, it gives excellent approximations. Error weighting functions can also improve the approximation, but their use requires solving the normal equations.

Time-domain window functions were shown to reduce the effects of truncation in the LS error FIR filter design procedure. Six different window functions were presented, and the results of the Hanning and Kaiser windows were shown in examples.

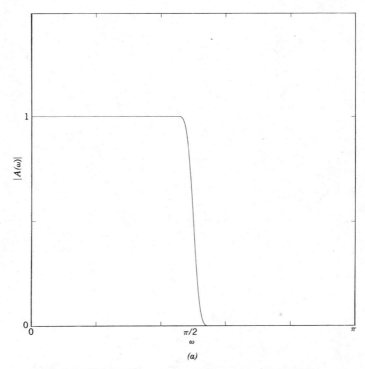

(a)

FIGURE 3.24. Length-101 low-pass filter with a Kaiser window.

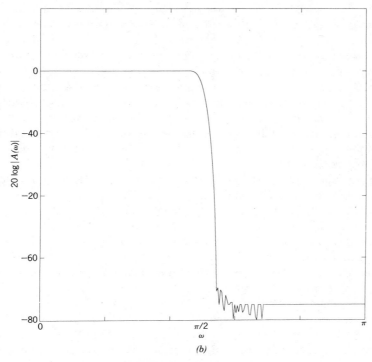

(b)

FIGURE 3.24. (Continued)

The use of windows is a somewhat ad hoc method of altering an optimal LS error design to reduce the Gibbs phenomenon at the band edge. This reduction of the overshoot increases the mean squared error but in a way difficult to predict. Because of this difficulty and the discovery, by Parks and McClellan,[1,2,31] that the Remes exchange algorithm directly attacks the overshoot reduction, windows are not used as much as they once were. If the simplicity of design by the windowed LS error method is desired, the Kaiser window is probably the best and certainly the most versatile.[1,2,4,9,10]

Program 4 in the appendix designs the six different types of windowed filters. As with all the design methods, their characteristics are best understood by experimentally designing filters with an interactive system where the frequency response can be easily displayed or plotted.

3.3 CHEBYSHEV APPROXIMATION

Figures 3.12 and 3.13 showed peaks or overshoots in the frequency response that are typical of frequency sampling and LS designs. The windowing techniques in Section 3.2.3.4 are attempts to reduce the peaks in the error function (the difference between the desired frequency response and the actual frequency response). By carefully applying various windows, the maximum error in the frequency response can be reduced. A natural question to ask is just how far can the maximum error be reduced? The theory of Chebyshev approximation, when applied to the filter design problem, answers this question and provides algorithms to find the coefficients of a linear-phase FIR filter that has a frequency response with this minimum value for the maximum error. An approximation that minimizes the maximum error over a set of frequencies is called a *Chebyshev approximation*.[17,18] Filters that have the minimum value of the maximum error exhibit an equiripple behavior in their frequency responses.[19] Thus, these optimum Chebyshev filters are sometimes called *equiripple* filters.

One of the earliest reports on the design of Chebyshev FIR digital filters was a General Electric report by M. A. Martin in 1962.[20] Most of the papers on the subject were published in the early 1970s. Tufts, Rorabacher, and Moser published some examples in 1970[21]; then Tufts and Francis[22] compared minimax designs with LS designs. Helms, in 1972[23], described techniques, including linear programming, to solve the Chebyshev approximation problem for filter design. In 1970, Herrmann published an article describing the equations that must be solved to obtain a filter with the maximum possible number of equal ripples[24] (Later called *extra-ripple*[25] or *maximal-ripple*[19] filters). Schüssler, in 1970, presented the work he and Herrmann had been doing on the design of maximal-ripple filters at the Arden House workshop.[26] Hofstetter developed an efficient algorithm for solving the equations proposed by Herrmann and Schüssler and presented papers with Oppenheim and Siegel at the 1971 Princeton conference[27] and the 1971 Allerton House conference[28] describing the algorithm and relating it to the Remes exchange algorithm.

In 1972, at the Arden House workshop, Parks described his work with McClellan on a direct application of Chebyshev approximation theory to the filter design problem using the Remes exchange algorithm,[29] and Parks and McClellan published a description of a design algorithm that used some of the computational techniques of Hofstetter's algorithm.[30] Hersey et al. described, at about the same time, an interactive method for designing filters with upper and lower constraints on the magnitude of the frequency response[31]. They also pointed out in their 1972 paper[31] that the Remes exchange algorithm could be used to design FIR linear-phase filters with the Chebyshev error criterion. The algorithm described in reference 30 has come to be known as the Parks–McClellan algorithm. A comprehensive paper, published with Rabiner,[19] gives a good summary of properties of filters designed with this algorithm. A program implementing the Parks–McClellan algorithm was published by the IEEE Press and is reprinted by permission of the IEEE as Program 6 in the appendix.

This section begins with a review of the characteristics of FIR filters with linear phase and describes the four different types of filters in detail. Some basic ideas from the theory of Chebyshev approximation are then presented. These concepts lead to the equiripple property of optimum filters. The Remes exchange algorithm is developed and adapted to the design of those linear-phase filters that best approximate a desired frequency characteristic in the Chebyshev sense.

3.3.1 Four Types of Linear Filters

As indicated in Section 2.2, there are four types of linear-phase FIR filters. All four types have a frequency response

$$H(f) = (j)^m A(f) e^{-j((N-1)/2)2\pi f}, \qquad m = 0, 1, \tag{3.43}$$

where $A(f)$ is a real-valued positive or negative function. In this section the frequency variable f, with units of cycles per second or hertz, is used along with a normalized unit sampling rate in order to be consistent with the literature in this area. The relation between this frequency variable and the radian frequency ω is $\omega = 2\pi f$. If the filter has $h(n) = h(N - 1 - n)$, it is said to have even symmetry and $m = 0$ in (3.43). If, on the other hand, $h(n) = -h(N - 1 - n)$, then the filter is said to have odd symmetry and $m = 1$ in (3.43). For even symmetry there are two types of filters corresponding to odd and even N. Similarly, for odd symmetry there are two additional types of filters for odd and even N. In reference 32 it is shown that $A(f)$ can always be written as a weighted sum of cosines for all four types of linear-phase filters. These formulas can be derived from equations (2.19), (2.21), and (2.24) with the use of appropriate trigonometric identities. The specific form of $A(f)$ is given in Table 3.1.

If the impulse response $h(n)$, $n = 0, \ldots, N - 1$, has an odd length (if N is odd),

TABLE 3.1. Approximating Functions for Linear-Phase Filters

	Symmetry	
	Even $h(n) = h(N-1-n)$, $(m = 0)$	Odd $h(n) = -h(N-1-n)$, $(m = 1)$
Odd Length (N)	$A(f) = \sum_{k=0}^{(N-1)/2} a_k \cos 2\pi kf$ $a_0 = h((N-1)/2)$ $a_k = 2h(-k + (N-1)/2)$ $k = 1, \ldots, (N-1)/2$	$A(f) = \sin 2\pi f \sum_{k=0}^{(N-3)/2} c_k \cos 2\pi kf$ $c_0 - \frac{1}{2}c(2) = 2h((N-3)/2)$ $c((N-5)/2) = 4h(1)$ $c((N-3)/2) = 4h(0)$ $c(k-1) - c(k+1) = 2h(-k + (N-1)/2)$ $k = 2, \ldots, (N-5)/2$
Even Length (N)	$A(f) = \cos \pi f \sum_{k=0}^{(N-3)/2} b_k \cos 2\pi kf$ $b_0 + \frac{1}{2}b(1) = 2h((N-3)/2$ $b((N-3)/2) = 4h(0)$ $b(k-1) + b(k) = 4h(-k + (N-1)/2)$ $k = 2, \ldots, (N-3)/2$	$A(f) = \sin \pi f \sum_{k=0}^{(N-3)/2} d_k \cos 2\pi kf$ $d_0 - \frac{1}{2}d(1) = 2h((N-3)/2)$ $d((N-3)/2) = 4h(0)$ $d(k-1) - d(k) = 4h(-k + (N-1)/2)$ $k = 2, \ldots, (N-3)/2$

there are two different linear-phase filters:

Type 1: even symmetry, odd length

Type 3: odd symmetry, odd length

As shown in Table 3.1 and Fig. 2.3, the odd-symmetry, odd-length filter (type 3) has a frequency response that must be zero at $f = 0$ and at $f = 0.5$. That is, a type 3 filter should not be used for either a low-pass or a high-pass design. Further, the type 3 filter introduces a phase shift of $90°$, as shown by (3.43).

If the impulse response $h(n)$ has an even length (if N is even), the resulting two linear-phase filters are

Type 2: even symmetry, even length

Type 4: odd symmetry, even length

As shown in Table 3.1 and Fig. 2.3, the odd-symmetry, even-length filter (type 4) has a frequency response that must be zero at $f = 0$ but not necessarily at $f = 0.5$. The type 4 filter will make a good highpass filter, but it should not be used for a low-pass filter. The even-symmetry, even-length filter (type 2) must be zero at $f = 0.5$ but not necessarily at $f = 0$. This filter type will make a good low-pass filter but not a good high-pass filter. As does the type 3 filter, the odd-symmetry, even-length filter (type 4) introduces a phase shift of $90°$, as shown by (3.43).

3.3.2 Chebyshev Approximation for Linear-Phase Design

The desired frequency response for an ideal low-pass filter is shown in Fig. 3.1a. The ideal response is real (no phase shift), exactly unity in the pass band, and exactly zero for the entire stop band. It is impossible for a causal FIR filter to have exactly zero phase (except for the trivial case when $h(0) = 1$ and all other coefficients are zero, in which case the stop-band transmission is unity). It is, however, possible to obtain an FIR filter with *linear* phase for all frequencies, as shown in Table 3.1. The group delay (negative of the derivative with respect to frequency of the phase function [16]) is a constant for all frequencies for linear-phase filters. Further, it is impossible for an FIR filter to have exactly zero transmission in the entire stop band (except for the trivial filter, which has zero transmission for all frequencies). An acceptable frequency response, shown in Fig. 3.1c, has the following characteristics:

1. Linear phase.
2. A width Δf transition band between the pass band and stop band.
3. A deviation from unity of $A(f)$ in the pass band of $\pm \delta_1$.
4. A deviation from zero of $A(f)$ in the stop band of $\pm \delta_2$.

A more general filter design problem could have several pass bands and several stop bands. Some of the bands could even consist of a single point when specifying, for example, that the filter have a transmission zero at a specific frequency. Further, some of these bands may be more important than others;

therefore different weights should be put on different bands. The multiple bands are assumed to make up a compact subset of the frequency band $[0, 0.5]$. The compact subset $\mathcal{F}$ in most applications is the union of closed intervals (corresponding to frequency bands) and discrete frequency points. These requirements for a good linear-phase filter are summarized in the following statement of the approximation problem for linear-phase design.

Approximation Problem for Linear-Phase Design
Given the following:

A compact subset $\mathcal{F}$ of $[0, 0.5]$.
A desired real-valued function $D(f)$, defined and continuous on $\mathcal{F}$.
A positive weight function $W(f)$, defined and continuous on $\mathcal{F}$.
The form of $A(f)$,

$$A(f) = Q(f) \sum_{k=0}^{r-1} c_k \cos(2\pi k f). \qquad (3.44)$$

We want to minimize over c_k

$$\|E(f)\| = \max_{F \in \mathcal{F}} W(f) \cdot |D(f) - A(f)|$$

by choice of $A(f)$.

Each of the four types of linear-phase filters in Table 3.1 is described by (3.44), where, by definition,

$$Q(f) = \begin{cases} 1, \\ \cos \pi f, \\ \sin 2\pi f, \\ \sin \rho f, \end{cases} \qquad (3.45)$$

After the coefficients of $A(f)$ are found, the impulse response of the filter can be determined from the simple relationships in Table 3.1.

The problem we have stated, which minimizes the maximum deviation over a set of frequencies, is the Chebyshev approximation problem for designing FIR filters. This problem leads directly to a characterization of the optimum filter in terms of the alternation theorem. Only the type 1 approximation with $Q(f)$ will be described in the following theory. The extension to the other three types of filters with their corresponding $Q(f)$ functions is straightforward.[19,33]

The alternation theorem states that there is a unique best Chebyshev approximation and that the (weighted) error of this optimum filter necessarily has an equiripple character.

Alternation Theorem
If $A(f)$ is a linear combination of r cosine functions—that is, if

$$A(f) = \sum_{k=0}^{r-1} c_k \cos(2\pi k f),$$

then a necessary and sufficient condition that $A(f)$ be the unique, best weighted Chebyshev approximation to a given continuous function $D(f)$ on $\mathscr{F}$ is that the weighted error function $E(f) = W(f) \cdot [D(f) - A(f)]$ exhibit *at least* $r + 1$ extremal frequencies in $\mathscr{F}$. These extremal frequencies are points such that, with $f_2 < f_2 < \cdots f_r < f_{r-1}$,

$$E(f_m) = -E(f_{m+1}) \qquad \text{for } m = 1, 2, \ldots, r \qquad (3.46)$$

and

$$|E(f_i)| = \max_{f \in \mathscr{F}} E(f). \qquad (3.47)$$

The alternation theorem means that the best Chebyshev approximation must necessarily have an equiripple error function. It also states that there is a *unique* best approximation for a given set of frequencies, filter length N, and weight function $W(f)$. The phrase "at least $r + 1$ extremal frequencies" needs some explanation. Since the best approximation for a given set of specifications is unique, there will not be one filter with $r + 1$ extremals and another filter with $r + 2$ extremals for the same specifications. For a given set of specifications, the unique best filter may have more than $r + 1$ extremal frequencies. If, for example, the optimum filter has $r + 3$ extremal frequencies, then by the uniqueness property, there cannot be a filter with only $r + 1$ extremals *for this set of specifications.*

In Fig. 3.25 the frequency response of an optimum length-13 linear-phase filter with even symmetry has eight extremal frequencies as required ($r = 7$). The function $A(f)$ for this filter is a sum of seven cosines (counting zero frequency). There is one more extremal frequency than there are degrees of freedom in $A(f)$, as required by the alternation theorem. The alternation theorem characterizes the optimum solution so that one can be recognized, but it does not directly show how to choose the filter coefficients. If the eight extremal frequencies were

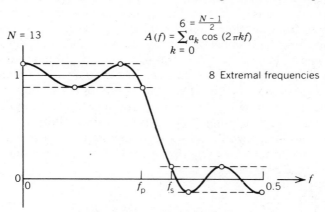

FIGURE 3.25. Frequency response for length-13 filter.

known, the impulse-response coefficients could be found easily by solving an interpolation problem by the frequency-sampling techniques in Section 3.1. In other words, if the extremal frequencies were used in the frequency-sampling design with desired values of $1.0 \pm \delta_1$ for the pass-band frequencies and $\pm \delta_2$ for the stop-band extremal frequencies, the impulse response of the optimum Chebyshev approximation filter would be obtained.

The problem of designing the filter has been reduced to the problem of finding the extremal frequencies. The Remes exchange algorithm[17,18] has proved to be valuable in finding these extremal frequencies.

3.3.3 The Remes Exchange Algorithm

The Remes exchange algorithm makes use of the fact that it is always possible to make the error function

$$E(f) = D(f) - \sum_{k=0}^{r-1} c_k \cos(2\pi k f) \tag{3.48}$$

take on the values $\pm \delta$ for any given set of $r + 1$ frequency points f_m, $m = 1, \ldots$, $r + 1$.[17,18] (To simplify notation, we assume a unit weight function, but these results apply to a general positive weight function.) In other words, the set of linear equations

$$D(f_m) = \sum_{k=0}^{r-1} c_k \cos(2\pi k f_m) + (-1)^m \delta, \qquad m = 1, \ldots, r + 1, \tag{3.49}$$

has a unique solution for the coefficients c_k, $k = 0, \ldots, r - 1$, and the amplitude. δ, of the oscillation on the given frequencies f_m. In the application of the Remes exchange algorithm developed by Parks and McClellan,[30] the set $\mathscr{F}$ of frequencies over which the approximation is made is an equally spaced grid with the number of frequency points approximately equal to 10 times the filter length. If $\mathscr{F}$ consisted only of the $r + 1$ frequencies f_m in (3.49), then the approximation problem would be solved in one step. The coefficients c_k in (3.49) would be the coefficients of the best approximation, and the maximum error on $\mathscr{F}$ would be $|\delta|$. This conclusion follows directly from the alternation theorem, where f_m are the extremal frequencies and δ is the amplitude of the oscillation. The error on one extremal frequency would be δ, and the error on the next extremal frequency would be $-\delta$. Furthermore, when $\mathscr{F}$ contains only $r + 1$ frequencies,

$$\max_{f \in \mathscr{F}} \left| D(f) - \sum_{k=0}^{r-1} c_k \cos(2\pi k f) \right| = |\delta|. \tag{3.50}$$

These are the conditions for f_m to be extremal frequencies.

In most practical applications $\mathscr{F}$ contains more than $r + 1$ frequencies. The problem in these cases is to find which subset of $r + 1$ frequencies is the set of extremal frequencies. The Remes exchange algorithm begins with a trial set of frequencies, as shown in Fig. 3.26, and systematically exchanges frequencies until the set of extremal frequencies is found. The new frequencies used in the next trial set are those $r + 1$ frequencies where the weighted error $E(f)$ has the largest magnitude.[30] Given a trial set of frequencies

$$T = \{f_1, f_2, \ldots, f_{r+1}\},$$

the Remes exchange algorithm consists of the following four basic computations:

1. Solve the linear equations in (3.49). This solution has an error that oscillates with amplitude δ_k on the trial set of frequencies for the kth iteration.
2. Interpolate to find the frequency response on the entire grid of frequencies.
3. Search over the entire grid of frequencies to see if (and where) the magnitude of the error in (3.48) is larger than the magnitude of δ_k found in step 1.
4. If the maximum value of the error magnitude found in step 3 equals δ_k, stop. If not, take the $r + 1$ frequencies where the error attains its maximum magnitude as the new trial set of extremal frequencies and go to step 1.

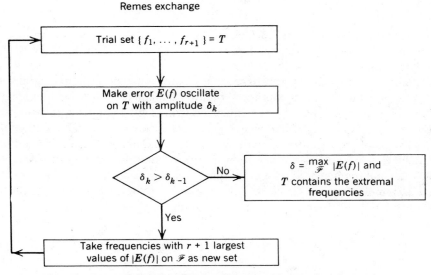

Remes exchange

δ_k increases on each iteration. The iteration stops when δ_k stops increasing. At this point, $\delta_k = \underset{\mathscr{F}}{\max} |E(f)|$ and T contains the $r + 1|$extremal frequencies.

FIGURE 3.26. Block diagram for Remes exchange.

It is often easier to think of the approximation problem in terms of polynomials. The frequency-domain approximation problem and the polynomial problem can be shown to be equivalent by using the change of variables

$$x = \cos(2\pi f) \tag{3.51}$$

or

$$f = \frac{\cos^{-1}(x)}{2\pi}. \tag{3.52}$$

With this change of variable, (3.44) becomes, with $Q(f) = 1$,

$$A\left(\frac{\cos^{-1} x}{2\pi}\right) = S(x) = \sum_{k=0}^{r-1} c_k \cos(k \cos^{-1}(x)) \tag{3.53}$$

or

$$S(x) = \sum_{k=0}^{r-1} d_k x^k. \tag{3.54}$$

The function $\cos(k \cos^{-1}(x))$ is indeed a polynomial.[17,18] The Chebyshev polynomials $C_k(x)$ have the form

$$C_k(x) = \cos(k \cos^{-1}(x)) \tag{3.55}$$

and are also used in Section 7.2.3.

The following simple example of the Remes exchange, with a first-order polynomial as the approximating function, illustrates the important features of this technique. For a more detailed description of the Remes exchange algorithm, see references 17 and 18.

Example 3.13. Remes Exchange
   The problem here is to choose the two coefficients d_0 and d_1 to minimize the Chebyshev error

$$\max_{x\in[0,1]} |x^2 - (d_0 + d_1 x)|.$$

In this problem a parabola is approximated by a straight line.
   Since two functions (the constant 1 and the function x) are being used in this approximation problem, there will be three extremal points. The trial set of extremal points is denoted by T, as in Fig. 3.27.
   The first, arbitrarily chosen, trial set is $T_0 = [0.25, 0.5, 1.0]$. To make the error

oscillate on these three points, we must solve the three linear equations

$$x_m^2 = d_0 + d_1 x_m + (-1)^m \delta, \qquad m = 0, 1, 2 \tag{3.56}$$

for δ and evaluate the error

$$E_0(x) = x^2 - (d_0 + d_1 x) \tag{3.57}$$

for all $x \in [0, 1]$ to see if there are any points where the error has a magnitude larger than $|\delta|$.

For this trial set of points the matrix version of the linear equations in (3.56) is

$$\begin{bmatrix} 1 & 0.25 & 1 \\ 1 & 0.5 & -1 \\ 1 & 1.0 & 1 \end{bmatrix} \begin{bmatrix} d_0 \\ d_1 \\ \delta_0 \end{bmatrix} = \begin{bmatrix} 0.0625 \\ 0.25 \\ 1.0 \end{bmatrix}. \tag{3.58}$$

The solution to these equations gives $\delta_0 = 0.0625$ and an error function shown in Fig. 3.27. Since the maximum value of the error on the interval [0, 1] is 0.3125, this trial set is not the extremal set (the error does not achieve its maximum magnitude on the trial set T_0).

The next trial set T_1 is made up of those three points in [0, 1] where the error $E_0(x)$ achieves its maximum magnitude. Thus,

$$T_1 = [0.0, 0.625, 1.0].$$

Again, the error is made to oscillate on this trial set by solving the linear equations

$$\begin{bmatrix} 1 & 0.0 & 1 \\ 1 & 0.625 & -1 \\ 1 & 1.0 & 1 \end{bmatrix} \begin{bmatrix} d_0 \\ d_1 \\ \delta_1 \end{bmatrix} = \begin{bmatrix} 0.0 \\ 0.390625 \\ 1.0 \end{bmatrix}. \tag{3.59}$$

The solution to these equations gives $\delta_1 = 0.1171875$ and an error function, shown in Fig. 3.27, with a maximum magnitude of 0.1328125. Since this maximum error is greater than δ_1, the trial set of points T_1 is not the extremal set. As shown in Fig. 3.27, the maximum error magnitude occurs at 0.0, 0.5, and 1.0. Thus, the next trial set is

$$T_2 = [0.0, 0.5, 1.0].$$

The error is made to oscillate on this new trial set of points by solving the linear equations

$$\begin{bmatrix} 1 & 0.0 & 1 \\ 1 & 0.5 & -1 \\ 1 & 1.0 & 1 \end{bmatrix} \begin{bmatrix} d_0 \\ d_1 \\ \delta_2 \end{bmatrix} = \begin{bmatrix} 0.0 \\ 0.25 \\ 1.0 \end{bmatrix}. \tag{3.60}$$

Choose d_0, d_1 to minimize $\max\limits_{x \epsilon [0,1]} |D(x) - (d_0 + d_1 x)|$

$D(x) = x^2$.

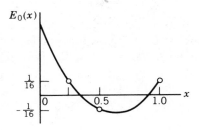

$T_0 = \left\{\tfrac{1}{4}, \tfrac{1}{2}, 1\right\}$ $\delta_0 = \tfrac{1}{16}$

$E_0 = x^2 - \tfrac{5}{4} x + \tfrac{5}{16}$, $||E_0|| = \tfrac{5}{16}$

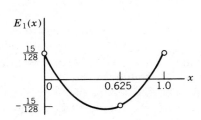

$T_1 = \left\{0, \tfrac{5}{8}, 1\right\}$ $\delta_1 = \tfrac{15}{128}$

$E_1 = x^2 - x + \tfrac{15}{128}$ $||E_1|| = \tfrac{17}{128}$

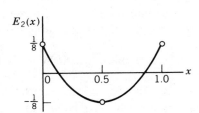

$T_2 = \left\{0, \tfrac{1}{2}, 1\right\}$ $\delta_2 = \tfrac{1}{8}$

$E_2 = x^2 - x + \tfrac{1}{8}$ $||E_2|| = \tfrac{1}{8}$

FIGURE 3.27. Example of Remes exchange.

This time $\delta_2 = 0.125$, *and* the maximum error is also equal to 0.125, as shown in Fig. 3.27. Thus, T_2 is the extremal point set because the error alternates in sign on these three points and achieves its maximum magnitude on each of these three points.

Example 3.13 illustrates the principal features of the Remes exchange. The error is made to oscillate on a trial set of points. New points where the error is larger than the amplitude of the oscillation are included (exchanged). Then the error is again forced to oscillate on this new set with a larger amplitude of oscillation. The amplitude of the oscillation, δ, increases at each iteration until it is equal to the maximum of the error over the entire interval of interest. At this point in the iterative algorithm, the points on which the error oscillates are the extremal points.

After we find the extremal points f_m, we can find the coefficients c_k in the approximation

$$A(f) = \sum_{k=0}^{r-1} c_k \cos(2\pi k f) \qquad (3.61)$$

by solving an interpolation problem of fitting the function $A(f)$ in (3.61) to the r known values. This procedure amounts to solving the following set of linear equations:

$$A(f_m) = \sum_{k=0}^{r-1} c_k \cos(2\pi k f_m) = D(f_m) \pm \delta, \qquad m = 1, \ldots, r. \qquad (3.62)$$

After we find c_k from (3.62), we easily calculate the impulse response values from Table 3.1.

The block diagram in Fig. 3.26 shows the Remes exchange technique used to design FIR digital filters. This figure is taken from reference 19 where the details of the implementation may be found. The FORTRAN program that implements the algorithm in Fig. 3.26 may be found in *Programs for Digital Signal Processing*.[9] A listing of a slightly modified version of this program is included in the appendix as Program 6. For a given set of filter specifications, the program formulates an equivalent approximation problem, which uses a weighted combination of cosines, as in (3.44), in a Chebyshev approximation problem. The Remes exchange is then used to find the extremal frequencies. When the extremal frequencies are found, the impulse response is found from the frequency response.

If a filter with a length of several hundred coefficients is needed, some modifications of Program 6 are necessary. Usually numerical problems will first occur in the interpolation step (step 2 in Section 3.3.3). These problems are especially likely to occur with very narrow pass bands and large transition bands. Bonzanigo[34] has developed an algorithm that can design filters with lengths in the thousands. The recent work of Ebert and Heute[35] should be valuable, especially when designing long filters. They have described several improvements to Program 6 that significantly reduce computing time.

3.3.4 Guidelines for Using the Parks–McClellan Algorithm

Although the filters that are designed by using the Parks–McClellan algorithm with the Remes exchange are indeed optimum in the sense that the maximum weighted error is minimized on the specified compact set of the frequency axis, they may not possess all of the characteristics desired. There are certain basic limitations to the performance of any of the four different types of linear-phase FIR filters, as described in Section 2.2.1. Complicated relations exist between the various parameters involved in the filter specification, such as the band-edge frequencies, the attenuations in the various bands, and the filter length. This section describes the characteristics of typical low-pass filters, using examples and empirical formulas relating the various parameters. The discussion is then extended from this two-band case (one pass band and one stop band) to the bandpass case with three bands.

Program 6 was used to calculate all of the examples in this section. SIG, the signal processing package developed by Lawrence Livermore Laboratory,[12] was used to draw the plots.

Example 3.14. Length-21 Low-pass Filter
In this design of a length-21 low-pass filter, all band-edge frequencies are given in fractions of the sampling frequency. A large pass band was used, with a frequency range from 0 to 0.33. The stop band was specified to begin at 0.37, and the errors in the pass band were given the same weight as errors in the stop band.

After the filter is designed, a summary of the resulting filter parameters is printed out, as shown in Fig. 3.28.

For this length-21 filter the amplitude function $A(\omega)$ is the sum of 11 cosine terms, and, according to the theory described in Section 3.3.1, the weighted error should have at least 12 extremal frequencies. The magnitude response in Fig. 3.29 exhibits these 12 extremal frequencies. The 12 points where the error achieves its maximum magnitude are circled in the figure. Notice that the two band edges 0.33 and 0.37 are extremal frequencies. This must always be the case.[33] In this example $f = 0.0$ and $f = 0.5$ are extremal frequencies; One of these two (either 0.0 or 0.5) must be always an extremal frequency, but it is not necessary that both be extremal frequencies.[33]

```
1***********************************************************
                finite impulse response (fir)
                linear phase digital filter design
                   remez exchange algorithm
                      bandpass filter

                   filter length =   21

              ***** impulse response *****
           h( 1) =   0.18255439e-01 = h( 21)
           h( 2) =   0.55136755e-01 = h( 20)
           h( 3) =  -0.40910728e-01 = h( 19)
           h( 4) =   0.14930855e-01 = h( 18)
           h( 5) =   0.27568584e-01 = h( 17)
           h( 6) =  -0.59407797e-01 = h( 16)
           h( 7) =   0.44841841e-01 = h( 15)
           h( 8) =   0.31902660e-01 = h( 14)
           h( 9) =  -0.14972545e+00 = h( 13)
           h(10) =   0.25687239e+00 = h( 12)
           h(11) =   0.69994062e+00 = h( 11)
                      band  1        band   2
lower band edge      0.            0.3700000
upper band edge      0.3300000     0.5000000
desired value        1.0000000     0.
weighting            1.0000000     1.0000000
deviation            0.0988697     0.0988697
deviation in db      0.8189238    -20.0987320
```

FIGURE 3.28. Filter parameters for length-21 low pass.

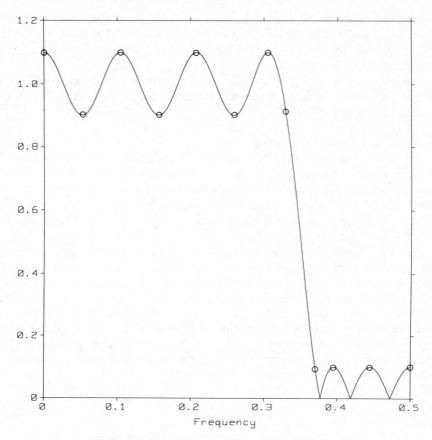

FIGURE 3.29. Magnitude response for length-21 low pass.

The unit-sample response for this example is shown in Fig. 3.30. This is a type 1 filter with an odd length and a positive symmetry. The frequency response is not forced to be zero at either $f = 0.0$ or at $f = 0.5$.

3.3.5 Design Formulas

For a low-pass filter the following five parameters are of interest:

N Filter length.
f_p The edge of the pass band specified as a fraction of the sampling frequency.
f_s The edge of the stop band specified as a fraction of the sampling frequency.
δ_1 The deviation from unity in the pass band.
δ_2 The deviation from zero in the stop band.

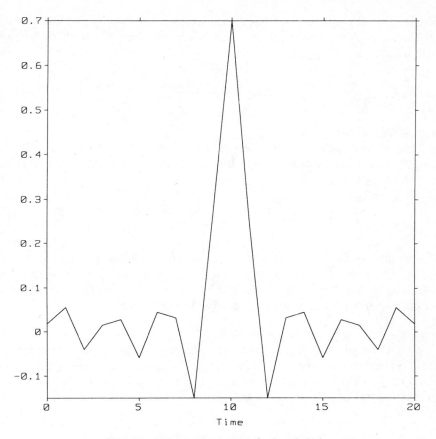

FIGURE 3.30. Unit-sample response for length-21 low pass.

Kaiser[19] has developed an empirical formula relating these parameters, using $\Delta f = f_s - f_p$ for the relative or normalized transition width.

$$N = \frac{-20 \log_{10}\sqrt{\delta_1 \delta_2} - 13}{14.6 \Delta f} + 1. \qquad (3.63)$$

When $\delta_1 = \delta_2$, (3.63) simplifies to

$$N = \frac{\text{dB} - 13}{14.6 \Delta f} + 1, \qquad (3.64)$$

where the stop-band attenuation in decibels is $\text{dB} = -20 \log_{10}\delta_2$.

Formula (3.63) gives a good initial value for the filter length N in most cases when the bandwidth is neither extremely wide nor extremely narrow. During the design of a very *narrow pass band*, the stop-band behavior governs the filter length (most of the frequency characteristic is stop band). Another empirical formula applies:

$$N = \frac{0.22 + dB/27}{\Delta f}, \qquad (3.65)$$

and, as before, $dB = -20 \log_{10}\delta_2$.

During the design of notch filters (or a low-pass filter with a very *wide pass band*), the pass-band ripple governs the filter length, and the empirical formula for filter length is

$$N = \frac{0.22 + (-20 \log_{10}\delta_1)/27}{\Delta f}. \qquad (3.66)$$

Formulas (3.63), (3.65), and (3.66) are easy to remember, and they provide a reasonable estimate for the filter length N. However, if a programmable calculator or a computer is available for estimating N, then the more accurate formulas provided in references 1 and 19 should be used.

Example 3.15. Length-20 Low-pass Filter

In this design of a length-20 low-pass filter, all band-edge frequencies are again given in fractions of the sampling frequency. The same band-edge frequencies as in Example 3.14 were used. The errors in the pass band were given the same weight as the errors in the stop band.

```
1*************************************************************
                finite impulse response (fir)
              linear phase digital filter design
                 remez exchange algorithm
                    bandpass filter

                  filter length =   20

              ***** impulse response *****
          h( 1) =   0.48411224e-01 = h( 20)
          h( 2) =   0.13537414e-01 = h( 19)
          h( 3) = -0.39344054e-01 = h( 18)
          h( 4) =   0.53151824e-01 = h( 17)
          h( 5) = -0.31608246e-01 = h( 16)
          h( 6) = -0.25162734e-01 = h( 15)
          h( 7) =   0.83330631e-01 = h( 14)
          h( 8) = -0.86372212e-01 = h( 13)
          h( 9) = -0.34074463e-01 = h( 12)
          h(10) =   0.56718868e+00 = h( 11)
                        band   1          band   2
lower band edge      0.              0.3700000
upper band edge      0.3300000       0.5000000
desired value        1.0000000       0.
weighting            1.0000000       1.0000000
deviation            0.0981161       0.0981161
deviation in db      0.8129656      -20.1651917
```

FIGURE 3.31. Length-20 low pass, Example 3.15.

A summary of the resulting filter parameters is shown in Fig. 3.31. The resulting error of 0.983 was slightly less than the error with the length-21 filter of Example 3.14. Even though fewer cosine terms are used for this length-20 filter (10 are used as shown in Table 3.1), the weighting term preceding the sum gives an extra stop-band zero. In Fig. 3.32 the filter has the necessary 11 extremal frequencies and the zero at $f = 0.5$ that always results with this type 2 filter (even length and positive symmetry). Both this filter and that in Example 3.14 have three stop-band zeros and eight extremal frequencies in the pass band. This example shows that a filter that is shorter by one coefficient may have a better response.

The impulse response, shown in Fig. 3.33, has two middle samples with the same value. That always happens with a type 2 filter because there is no central sample. A delay of 9.5 samples rather than 10.0 in Example 3.14, occurs with this filter. The half-sample delay for even-length filters must be taken into account when using this filter in a system.

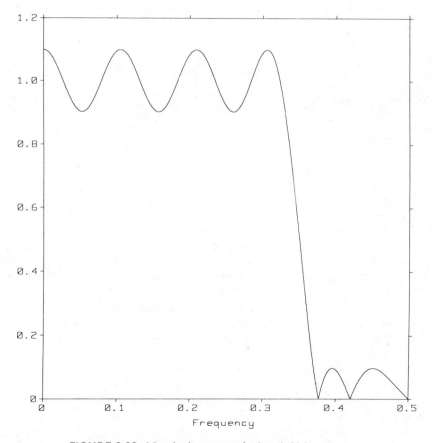

FIGURE 3.32. Magnitude response for length-20 low pass.

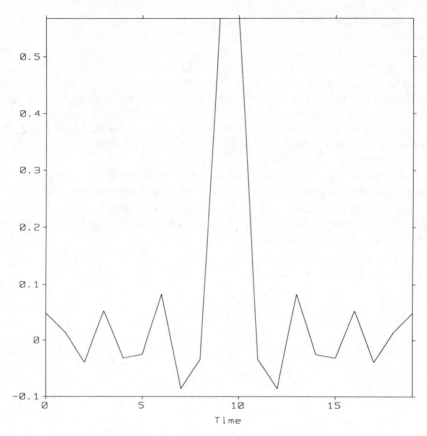

FIGURE 3.33. Unit-sample response for length-20 low pass.

Example 3.16. Low-pass Filter with Echoes

The low-pass filter with parameters shown in Fig. 3.34 and unit-sample response shown in Fig. 3.35 indicate the possibility of echoes in the pulse response of the filter. This filter has a sharp cutoff (a narrow transition band) and a large amplitude ripple in the pass band. As shown in reference 16, the large pass-band ripple results in side lobes or echoes in the impulse response. The amplitude of these echoes is directly proportional to the amplitude of the pass-band ripple. For this example the echoes are located at 2 samples and at 52 samples after the beginning of the impulse response. These echoes can be eliminated by redesigning the filter with a wider transition band and, therefore, a smaller pass-band ripple (assuming that the same 40 dB of attenuation is needed in the stop band). However, if a transition band as narrow as this one is needed, then the unique best Chebyshev approximation must have echoes in the impulse response. Filters designed by windowing methods usually do not have such large echoes. Most windows are small near the ends, thus attenuating the impulse

```
1**********************************************************A
            finite impulse response (fir)
            linear phase digital filter design
               remez exchange algorithm
                   bandpass filter

                  filter length =  55

              ***** impulse response *****
          h( 1) = -0.97759655e-02 = h( 55)
          h( 2) =  0.24792071e-01 = h( 54)
          h( 3) =  0.65477327e-01 = h( 53)
          h( 4) =  0.23820495e-01 = h( 52)
          h( 5) = -0.21147884e-01 = h( 51)
          h( 6) =  0.89294007e-02 = h( 50)
          h( 7) =  0.99083073e-02 = h( 49)
          h( 8) = -0.17189449e-01 = h( 48)
          h( 9) =  0.86856764e-02 = h( 47)
          h(10) =  0.79506608e-02 = h( 46)
          h(11) = -0.17969904e-01 = h( 45)
          h(12) =  0.11324050e-01 = h( 44)
          h(13) =  0.74154064e-02 = h( 43)
          h(14) = -0.21140970e-01 = h( 42)
          h(15) =  0.15478048e-01 = h( 41)
          h(16) =  0.71804966e-02 = h( 40)
          h(17) = -0.26794923e-01 = h( 39)
          h(18) =  0.22209013e-01 = h( 38)
          h(19) =  0.72455141e-02 = h( 37)
          h(20) = -0.36636721e-01 = h( 36)
          h(21) =  0.34442522e-01 = h( 35)
          h(22) =  0.73146555e-02 = h( 34)
          h(23) = -0.57848476e-01 = h( 33)
          h(24) =  0.64485028e-01 = h( 32)
          h(25) =  0.73688934e-02 = h( 31)
          h(26) = -0.14114894e+00 = h( 30)
          h(27) =  0.27177680e+00 = h( 29)
          h(28) =  0.67406112e+00 = h( 28)
                        band  1          band  2
lower band edge       0.               0.3500000
upper band edge       0.3300000        0.5000000
desired value         1.0000000        0.
weighting             1.0000000       20.0000000
deviation             0.1863634        0.0093182
deviation in db       1.4843543      -40.6133842
```

FIGURE 3.34. Parameters for low-pass with echoes.

response at the ends. A filter designed with windowing cannot, for the same stop-band attenuation and the same transition width, have any smaller pass-band error than this filter. The shape of the pass band will be different, and the maximum deviation from unity will be larger than for the unique optimum filter unless the window design *is* the best Chebyshev approximation with the required number of extremal frequencies.

Example 3.17. A Length-21 Bandpass Filter

In this example a length-21 filter is designed with three bands: one pass band and two stop bands. The errors in all three bands are weighted in the same way, as shown in Fig. 3.36. The two transition bands for this bandpass filter have the same width, 0.04. This width is identical to that of the low-pass filters in Examples 3.14 and 3.15. The error here is only slightly larger than in the low-

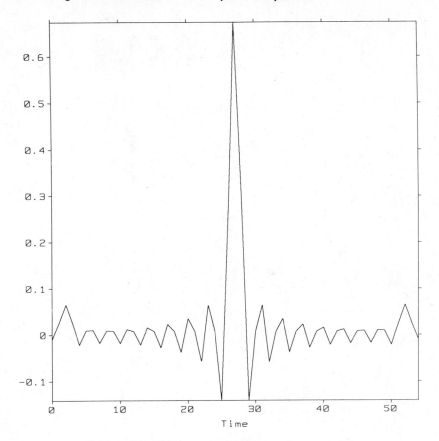

FIGURE 3.35. Unit-sample response for filter with echoes.

pass case. The magnitude response in Fig. 3.37 shows that the error has 12 extremal frequencies as required. Actually, the alternation theorem states that the length-21 filter must have *at least* 12 extremal frequencies. Reference 19 shows that it is possible to have more than 12.

This filter behaves as expected. However, the frequency response has only seven places, excluding $f = 0.0$ and $f = 0.5$, where the derivative is zero. It is possible to have a zero derivative at nine frequencies. This zero derivative and resulting local maximum (or minimum) may occur in the transition band and lead to unexpected results, as illustrated in Example 3.18.

Example 3.18. A Bandpass Filter with Transition Band Peak

We obtained this example by slightly modifying the specifications for the length-21 bandpass filter in Example 3.17. The first transition band was widened from 0.04 to 0.17, and the second transition was reduced from 0.04 to 0.03. The resulting filter parameters (Fig. 3.38) indicate that a reasonable design has been obtained. The error is slightly larger than with both transition bands equal to

```
1*****************************************************************
                finite impulse response (fir)
              linear phase digital filter design
                  remez exchange algorithm
                    bandpass filter

                  filter length =   21

                ***** impulse response *****
          h( 1) =   0.46677580e-02 = h( 21)
          h( 2) =   0.96759470e-02 = h( 20)
          h( 3) = -0.90181328e-01 = h( 19)
          h( 4) = -0.25750486e-01 = h( 18)
          h( 5) =   0.45590442e-01 = h( 17)
          h( 6) = -0.10308806e-01 = h( 16)
          h( 7) =   0.11038479e+00 = h( 15)
          h( 8) =   0.12596332e-01 = h( 14)
          h( 9) = -0.28589711e+00 = h( 13)
          h(10) = -0.17343471e-01 = h( 12)
          h(11) =   0.38577724e+00 = h( 11)
```

	band 1	band 2	band 3
lower band edge	0.	0.1800000	0.3700000
upper band edge	0.1400000	0.3300000	0.5000000
desired value	0.	1.0000000	0.
weighting	1.0000000	1.0000000	1.0000000
deviation	0.1073546	0.1073546	0.1073546
deviation in db	-19.3835869	0.8857346	-19.3835869

FIGURE 3.36. Bandpass filter parameters.

0.04, but otherwise the filter seems to be a good bandpass filter. The unit-pulse response in Fig. 3.38 gives the unique best Chebyshev approximation *on the frequency bands specified.* On two transition bands there is no control of the error. These "don't care" regions should always be checked to verify that the frequency response has the expected monotonic behavior.

In this example, as in Example 3.16, a local maximum (or minimum) in a transition band is possible because not all of the places where the derivative is zero occur in the specified bands. Figure 3.39 shows that there is a local maximum in the first transition band, which gives a filter with quite different characteristics than expected. This phenomenon has been studied,[36] and some suggestions have been made to reduce the possibility of transition band peaks. The method recommended in reference 36 makes use of (3.63), the formula relating the transition width Δf and the errors δ_1 and δ_2. If there are two transition bands, Δf_1 and Δf_2, as in Example 3.18, with δ_1 and δ_2 the deviations on both sides of transition Δf_1 and with δ_2 and δ_3 the deviations on both sides of transition Δf_2, we first calculate

$$N_1 = \frac{-20\log_{10}(\sqrt{\delta_1 \delta_2}) - 13}{14.6\,\Delta f_1} + 1$$

and

$$N_2 = \frac{20\log_{10}(\sqrt{\delta_2 \delta_3}) - 13}{14.6\Delta f_2} + 1.$$

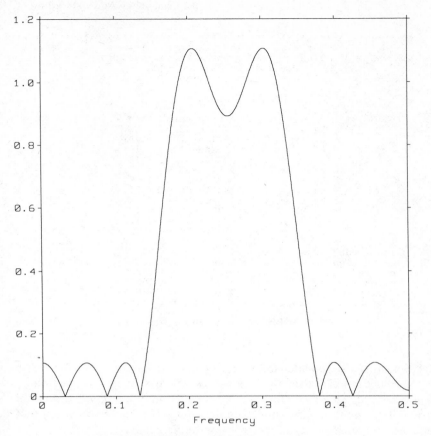

FIGURE 3.37. Magnitude response for bandpass filter.

```
1*********************************************************
                finite impulse response (fir)
                linear phase digital filter design
                    remes exchange algorithm
                      bandpass filter

                  filter length =  21

             ***** impulse response *****
          h( 1) = -0.21529701e+00 = h( 21)
          h( 2) = -0.22953761e+00 = h( 20)
          h( 3) =  0.54982297e-01 = h( 19)
          h( 4) =  0.44166148e+00 = h( 18)
          h( 5) =  0.69435006e+00 = h( 17)
          h( 6) =  0.11394271e+00 = h( 16)
          h( 7) = -0.71542710e+00 = h( 15)
          h( 8) = -0.10141391e+01 = h( 14)
          h( 9) = -0.71523160e+00 = h( 13)
          h(10) =  0.66042960e+00 = h( 12)
          h(11) =  0.17381012e+01 = h( 11)
```

	band 1	band 2	band 3
lower band edge	0.	0.2500000	0.4000000
upper band edge	0.0800000	0.3700000	0.5000000
desired value	0.	1.0000000	0.
weighting	1.0000000	1.0000000	1.0000000
deviation	0.1104313	0.1104313	0.1104313
deviation in db	-19.1381569	0.9098340	-19.1381569

FIGURE 3.38. Parameters for transition peak example.

One of these quantities will be larger than the other and will be the major factor in determining filter length. If N_2 is larger than N_1, as in Example 3.18, the "modified stop-band" method[36] suggests reducing the width Δf_1 by moving the stop-band edge frequency closer to the pass-band edge until N_1 and N_2 are approximately equal. Another approach, which does not allow exact specification of band-edge frequencies, is to ensure that all possible zeros of the derivative occur in the specified bands (maximal-ripple filter). Alternatively, rather than a transition band where there is no control of the error, a band can be used with a small weight on the error. There are other approaches that always put constraints on all frequencies and thereby guarantee that there will be no unexpected results in transition bands.[31,37,38] When linear programming is used to design the filter, an upper constraint on the response in the transition band can easily be specified.[39] When the Parks–McClellan algorithm is used for bandpass filters, the frequency response should always be examined in the transition bands, especially when there is a big variation in the widths of the transition bands, as in Example 3.18.

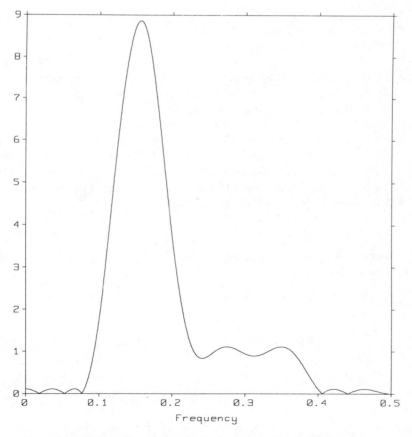

FIGURE 3.39. Overall frequency response with transition peak.

Summary

Filters that minimize the Chebyshev error measure necessarily have an equiripple characteristic for the weighted error function. The unique best Cheyshev filter must have at least one more extremal frequency (ripple) than the number of degrees of freedom in the impulse response: $N/2$ for even length N, and $(N + 1)/2$ for odd N.

The Remes exchange used in the Parks–McClellan algorithm provides an efficient method for designing Chebyshev equiripple filters. Filters can be designed with several pass bands and stop bands with different weighting in each band by using Program 6 in the appendix. This FORTRAN program can be modified easily to accommodate an arbitrary desired function and an arbitrary weight function.

This section provided several examples to illustrate special characteristics of Chebyshev equiripple filters and to indicate possible pitfalls in designing these filters.

3.4 DESIGN OF MAXIMALLY FLAT (BUTTERWORTH) FILTERS

When the filter tolerance scheme is stated in terms of the maximum allowable deviation from unity in the pass band and the maximum allowable deviation from zero in the stop band, the optimum filter has an equiripple magnitude characteristic (see Section 3.3). This equiripple characteristic can lead to "echoes" in the impulse response, as shown by Example 3.16, and is thus not always the most desirable magnitude shape. A smoother frequency magnitude characteristic leads to an impulse response with smaller amplitudes in the tails of the impulse response. The maximally flat, or Butterworth, FIR filter may be a useful alternative to the equiripple designs of Section 3.3.

3.4.1 Derivation of the Maximally Flat Linear-Phase Low-Pass Filter

The frequency response of a type 1 linear-phase filter has the form

$$H(f) = e^{-j2\pi f(N-1)/2} \sum_{k=0}^{(N-1)/2} d_k \cos(2\pi k f) \qquad (3.67)$$

or

$$H(f) = e^{-j2\pi f(N-1)/2} G(f). \qquad (3.68)$$

The coefficients in (3.67) are chosen to satisfy the following conditions:

$$H(f)\bigg|_{f=0} = 1.0, \qquad\qquad (3.69)$$

$$\frac{d^m}{df^m} H(f)\bigg|_{f=0} = 0.0, \qquad m = 1, 2, \ldots, N - k, \qquad (3.70)$$

$$\frac{d^m}{df^m} H(f)\bigg|_{f=0.5} = 0.0, \qquad m = 0, 1, \ldots, k - 1, \qquad (3.71)$$

where k is a constant chosen to give the desired amount of flatness at $f = 0$ and $f = 0.5$.

With the change of variable $x = \cos(2\pi f)$, described in Section 3.3.3, $G(f)$ in (3.68) becomes

$$P(x) = \sum_{k=0}^{(N-1)/2} d_k x^k. \qquad\qquad (3.72)$$

Condition (3.70) requires that $P(x)$ have a zero of order k at $x = 1$, and condition (3.71) requires that $P(x) - 1$ have a zero of order $(N + 1)/2 - k$ at $x = 0$. Herrmann[40] has given the explicit, closed-form expression for a polynomial satisfying these conditions:

$$P(x) = (1 - x)^k \sum_{m=0}^{(N-1)/2-k} \binom{k + m - 1}{m} x^m, \qquad (3.73)$$

where $\binom{k+m-1}{m}$ represents the binomial coefficient.

In applications using (3.73) for a low-pass filter, the pass-band edge cannot be specified exactly. The parameter k can be used to indirectly control the location of the band edge. A large value of k gives a smooth frequency response at zero frequency and gives a wide-band filter. A small value of k puts more emphasis on smoothness at $f = 0.5$ and thus gives a narrower pass band. Herrmann[40] has related k to the desired half-power, or cutoff, point $x = x_0$, using

$$k = \frac{N-1}{2} - \left[\frac{N-1}{2} \cdot x_0 + 0.5\right], \qquad (3.74)$$

where $[y]$ is the greatest integer less than y.

After the polynomial $P(x)$ is calculated, the impulse response may be calculated by interpolating $G(f)$ to the frequency response $P(\cos(2\pi f))$ in a manner similar to that used in the Parks–McClellan implementation of the Remes exchange. A program, written by J. F. Kaiser, that implements the foregoing procedure may be found in reference 9.

3.4.2 Smooth Pass Bands and Equiripple Stop Bands

Recently, a new class of FIR linear-phase bandpass filters with smooth pass bands has been described.[41] These filters have all their zeros on the unit circle with an equiripple stop-band behavior. The design of these filters is quite fast since, as in the case of the maximally flat filters, there is a closed-form expression for the response. These formulas are based on Zolatarev polynomials and are described in detail in reference 41. The Chebyshev polynomials described in Section 3.3.2 are a special case of the Zolotarev polynomials. These Zolotarev filters combine the smooth pass band of the maximally flat filters with the equiripple response of the Chebyshev filters. They are the FIR version of Chebyshev type 2 or inverse Chebyshev IIR filters described in Chapter 7.

Summary

This section described an alternative to the equiripple frequency characteristic. The maximally flat designs do not have as small deviations from the ideal response of unity in the pass band and zero in the stop band as do the Chebyshev designs, but they do have a smoother frequency response and an impulse response that has smaller amplitude tails than the Chebyshev designs. There are FIR filters that combine the Chebyshev-type equiripple stop band with a smooth Butterworth-type pass band based on Zolotarev polynomials.

REFERENCES

[1]   L. R. Rabiner and B. Gold, *Theory and Application of Digital Signal Processing*, Englewood Cliffs, NJ: Prentice-Hall, 1975.

[2]   A. V. Oppenheim and R. W. Schafter, *Digital Signal Processing*, Englewood Cliffs, NJ: Prentice-Hall, 1975.

[3]   F. J. Taylor, *Digital Filter Design Handbook*, New York: Dekker, 1983.

[4]   L. B. Jackson, *Digital Filters and Signal Processing*, Boston: Kluwer, 1986

[5]   L. R. Rabiner and C. M. Rader, eds., *Digital Signal Processing*, selected reprints, New York: IEEE Press, 1972.

[6]   *Digital Signal Processing II*, selected reprints, New York: IEEE Press, 1979. Edited by the Digital Signal Processing Committee.

[7]   J. E. Dennis, Jr. and R. B. Schnabel, *Numerical Methods for Unconstrained Optimization and Nonlinear Equations*, Englewood Cliffs, NJ: Prentice-Hall, 1983.

[8]   J. J. Dongarra, J. R. Bunch, C. B. Moler, and G. W. Stewart, *LINPACK Users' Guide*, Philadelphia: SIAM, 1979.

[9]   *Programs for Digital Signal Processing*, New York: IEEE Press, 1979.

[10]   *Digital Filter Design Package, DFDP*, Interactive Software for Digital Filter Design, Version 1.02, Atlanta, GA: Atlanta Signal Processors Inc., 1984.

[11] J. O'Donnell, *DISPRO v1.0 User's Manual*, Digital Filter Design Software, Wayland, MA: Signix Corp., 1983.

[12] *SIG: A General Purpose Signal Processing, Analysis, and Display Program*, Livermore, CA: Lawrence Livermore Labs, 1985.

[13] C. S. Burrus and T. W. Parks, *DFT/FFT and Convolution Algorithms*, New York: Wiley-Interscience, 1985.

[14] D. F. Elliot and K. R. Rao, *Fast Transforms: Algorithms, Analysis and Applications*, New York: Academic Press, 1982.

[15] R. Bracewell, *Fourier Transforms*, New York: McGraw-Hill, 1975.

[16] A. Papoulis, *The Fourier Integral and Its Applications*, New York: McGraw-Hill, 1962.

[17] E. W. Cheney, *Introduction to Approximation Theory*, New York: McGraw-Hill, 1966.

[18] J. R. Rice, *The Approximation of Functions*, Vol. 1, Reading, MA: Addison-Wesley, 1964.

[19] L. R. Rabiner, J. H. McClellan, and T. W. Parks, "FIR Digital Filter Design Techniques Using Weighted Chebyshev Approximation," *Proc. IEEE* **63**, 595–610 (1975).

[20] M. A. Martin, *Digital Filters for Data Processing*, Tech. Report No. 62-SD484, Missile and Space Division, General Electric Co., 1962.

[21] D. W. Tufts, D. W. Rorabacher, and M. E. Mosier, "Designing Simple, Effective Digital Filters," *IEEE Trans. Audio Electroacoustics* **AU-18**, 142–158 (1970).

[22] D. W. Tufts and J. T. Francis, "Designing Digital Low-Pass Filters—Comparison of Some Methods and Criteria," *IEEE Trans. Audio Electroacoustics* **AU-18**, 487–494 (1970).

[23] H. D. Helms, "Digital Filters with Equiripple or Minimax Responses," *IEEE Trans. Audio Electroacoustics* **AU-19**, 87–94 (1971). (Reprinted in reference 5).

[24] O. Herrmann, "Design of Nonrecursive Digital Filters with Linear Phase," *Electronics Lett.* **6**, 328–329 (1970). (Reprinted in ref. 4).

[25] T. W. Parks, L. R. Rabiner, and J. H. McClellan, "On the Transition Width of Finite Impulse-Response Digital Filters," *IEEE Trans. Audio Electroacoustics* **AU-21**, 1–4 (1973).

[26] O. Herrmann and H. W. Schüssler, *On the Design of Selective Nonrecursive Digital Filters*, presented at the IEEE Arden House Workshop on Digital Filtering, January 12, 1970.

[27] E. M. Hofstetter, A. V. Oppenheim, and J. Siegel, "A New Technique for the Design of Non-Recursive Digital Filters," in *Proceedings of the Fifth Annual Princeton Conference on Information Sciences and Systems*, pp. 64–72, 1971. (Reprinted in reference 4).

[28] E. M. Hofstetter, A. V. Oppenheim, and J. Siegel, "On Optimum Nonrecursive Digital Filters," *Proceedings of the Ninth Annual Allerton Conference on Circuit and System Theory*, pp. 789–798, 1971. 'Reprinted in Digital Signal Processing II, New York: IEEE Press. (1972).

[29] T. W. Parks, *Extensions of Chebyshev Approximation for Finite Impulse Response Filters*, presented at the IEEE Arden House Workshop on Digital Filtering, January 10, 1972.

[30]   T. W. Parks and J. H. McClellan, "Chebyshev Approximation for Nonrecursive Digital Filters with Linear Phase," *IEEE Trans. Circuit Theory* **CT-19**, 189–194 (1972).

[31]   H. S. Hersey, D. W. Tufts, and J. T. Lewis, "Interactive Minimax Design of Linear Phase Nonrecursive Digital Filters Subject to Upper and Lower Function Constraints," *IEEE Trans. Audio Electroacoustics* **AU-20**, 171–173 (1972).

[32]   J. H. McClellan and T. W. Parks, "A Unified Approach to the Design of Optimum FIR Linear-Phase Digital Filters," *IEEE Trans. Circuits Systems* **CT-20**, 697–701 (1973).

[33]   J. H. McClellan, *On the Design of One-Dimensional and Two-Dimensional FIR Digital Filters*, Ph.D. dissertation, Rice University, Houston, TX, 1973.

[34]   F. Bonzanigo, Private communication, ETH Zurich.

[35]   S. Ebert and U. Heute, "Accelerated Design of Linear or Minimum-Phase FIR Filters with a Chebyshev Magnitude Response," *Proc. IEE (British)* **130**, 267–270 (1983).

[36]   L. R. Rabiner, J. F. Kaiser, and R. W. Schafer, "Some Considerations in the Design of Multiband Finite-Impulse-Response Digital Filters," *IEEE Trans. Audio Electroacoustics* **ASSP-22**, 462–472 (1974).

[37]   F. Grenez, *Constrained Chebyshev Approximation for FIR Filters*, ICASSP-83, Boston, 1983, pp. 194–196.

[38]   M. T. McCallig and B. J. Leon, "Constrained Ripple Design of FIR Digital Filters," *IEEE Trans. Circuits Systems* **CAS-25**, 893–902 (1978).

[39]   K. Steiglitz and T. W. Parks *What Is the Filter Design Problem?* Twentieth Princeton Conference, Princeton, NJ, March 1986.

[40]   O. Herrmann, "On the Approximation Problem in Nonrecursive Digital Filter Design," *IEEE Trans. Circuit Theory* **CT-18**, 411–413 (1971).

[41]   X. Chen and T. W. Parks, "Analytic Design of Optimal FIR Smooth Passband Filters Using Zolotarev Polynomials," *IEEE Trans. Circuits Systems*, November CAS-33, 1065–1071 (1986).

4
Minimum-Phase and Complex Approximation

The advantage of linear-phase filters, as discussed in Chapter 2, is that the group delay is a constant for all frequencies. In other words, there is no delay distortion for linear-phase filters. The problem with linear-phase filters, however, is that this constant delay is always equal to $(N - 1)/2$, where N is the filter length. When a large attenuation is required in the stop band and a sharp cutoff is desired, N must be quite large (see Section 3.3). Thus, linear-phase filters with large stop-band attenuation and a sharp cutoff must have a large, but constant, delay. This large delay could be a major drawback of a filter, for it could cause instability if the filter were inside a feedback loop in a digital control system, difficulties in a telephone network (such as delays when using a satellite link), or the loss of large sums of money when trying to predict cycles in the stock market.

In many applications the ideal filter would have a large stop-band attenuation, a sharp cutoff, and zero phase shift (zero delay). However, such a filter is mathematically impossible. When a filter with less delay is desired, the minimum-phase filter is a good choice. Minimum-phase filters have all of their zeros inside or possibly on the unit circle. A minimum-phase filter can be obtained from a linear-phase filter by reflecting all of the zeros that are outside the unit circle to the inside of the unit circle. In other words, those zeros located at $z = re^{j\theta}(r > 1)$ are changed to zeros located at $z = e^{-j\theta}/r$. The resulting modified linear-phase filter will have minimum phase and will have the same magnitude (except for a scale factor) as the linear-phase filter. However, the minimum-phase filter obtained from the linear-phase filter in this way may not have the best possible magnitude characteristic.

This chapter discusses minimum-phase filter design in detail. The basic approach is essentially the same as in reference 1. Although other methods have been suggested,[2,3] none are clearly superior to the more direct approach of reference 1. The optimum-magnitude characteristic for a minimum-phase filter is characterized in terms of a minimum-phase alternation theorem. Minimum-phase filters with desirable magnitude characteristics are designed and compared with their linear-phase counterparts in terms of delay and magnitude characteristics.

Although the minimum-phase filter has a smaller group delay (minimum delay), the delay is not a constant for all frequencies, as it is for linear-phase filters. Another alternative, in addition to the linear- and minimum-phase designs, is the direct design with a complex desired function. A direct complex approximation is also required when the phase must be specified, as in the design of equalizers. In this chapter the complex approximation problem is formulated in such a way that linear programming may be used for the design. An example with approximately constant delay, which is less than the delay resulting from linear-phase design, is given. Complex approximation is also applied to the design of FIR equalizers.

4.1 OPTIMUM-MAGNITUDE CHEBYSHEV DESIGN

A length-N FIR filter with unit-pulse response $h_0, h_1, \ldots, h_{N-1}$ has a frequency response

$$H(f) = \sum_{n=0}^{N-1} h_n e^{-jn2\pi f}. \tag{4.1}$$

The squared magnitude of the frequency response in (4.1) is

$$|H(f)|^2 = \sum_{n=0}^{N-1} a_n \cos(n2\pi f), \tag{4.2}$$

where the a_n coefficients depend on the unit-pulse response values, h_n.

One's initial reaction to (4.2) is to try using the programs already developed for linear-phase design to design filters with a desirable squared magnitude, since these programs work with sums of cosines just like (4.2). A major stumbling block in such an approach is the complicated nonlinear relationship between the cosine coefficients (the a_n's) in (4.2) and the unit-pulse response of the corresponding filter [the h_n's in (4.1)]. A simple length-2 example illustrates this nonlinear relationship.

Example 4.1 A Length-2 Magnitude Characteristic
   When $N = 2$, the squared magnitude of the filter's frequency response is

$$|H(f)|^2 = H(f) \cdot H(f)^* \tag{4.3}$$

or

$$|H(f)|^2 = (h_0 + h_1 e^{-j2\pi f})(h_0 + h_1 e^{+j2\pi f}) \tag{4.4}$$

or

$$|H(f)|^2 = h_0^2 + h_1^2 + 2h_0h_1 \cos(2\pi f). \tag{4.5}$$

In other words, the nonlinear relation between the h's and the a's in (4.1) and (4.2) is, in this example,

$$a_0 = h_0^2 + h_1^2 \quad \text{and} \quad a_1 = 2h_0h_1. \tag{4.6}$$

For longer filters the nonlinear relationships between coefficients become far too complicated to solve easily for the h's. An alternative procedure that requires factoring a polynomial (also a nonlinear operation) is described in Section 4.1.2.

4.1.1 Characterization of Optimum-Magnitude Filters

The optimum-magnitude response has an equiripple characteristic as outlined in what follows. The squared magnitude response, shown in Fig. 4.1, is optimum in the sense described in the alternation theorem for the minimum-phase case, given next.[4]

Minimum-Phase Alternation Theorem
Given $K = \delta_1/\delta_2$, δ_2 is minimum if and only if $D(f) - |H(f)|^2 = E(f)$ has at least $N + 1$ extremal frequencies on B, where

N is the filter length = (number of coefficients)
δ_1 is the pass-band deviation
δ_2 is the stop-band deviation
B_p is the set of pass-band frequencies
B_s is the set of stop-band frequencies

$$B = [0, f_p] \cup [f_s, 0.5] = B_p \cup B_s, \tag{4.7}$$

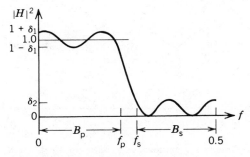

FIGURE 4.1. Optimum squared magnitude.

and

$$D(f) = \begin{cases} 1, & f \in B_p, \\ \dfrac{\delta_2}{2}, & f \in B_s. \end{cases} \tag{4.8}$$

This theorem can be used to identify an optimum-magnitude response. For example, a length-7 linear-phase (type 1) filter has a frequency response

$$H(f) = e^{-j6\pi f} A(f), \tag{4.9}$$

where the amplitude, as described in Section 2.2, is

$$A(f) = \sum_{k=0}^{3} b_k \cos(k2\pi f). \tag{4.10}$$

The alternation theorem in Section 3.3 states that $A(f)$ must have at least five extremal frequencies, as shown in Fig. 4.2a. The squared magnitude of the

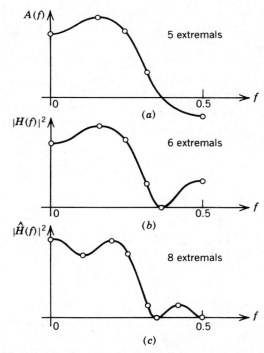

FIGURE 4.2. Linear-phase versus optimum magnitude. (*a*) Amplitude for length-7 linear-phase filter; (*b*) squared magnitude for filter in (*a*); and (*c*) optimum squared magnitude for length-7.

frequency response

$$|H(f)|^2 = |A(f)|^2 \tag{4.11}$$

is shown in Fig. 4.2b. The error function $E(f)$ described in the minimum-phase alternation theorem has only six extremal frequencies, as indicated in the figure. Even though this filter has an equiripple magnitude characteristic, it does not have the best possible magnitude characteristic in the sense of the minimum-phase alternation theorem. The best possible magnitude characteristic for a length-7 filter would have at least *eight* extremals, as illustrated in Fig. 4.2c. The filter with this optimum magnitude would not, however, have the desirable linear-phase characteristic. Instead it would be designed to have all of its zeros either inside the unit circle or possibly on the unit circle—that is, to have minimum phase.

4.1.2 Design Procedure

The design procedure described here was first proposed by Herrmann and Schüssler[1] and involves factoring a polynomial. There are several other approaches,[2,3] but this procedure is easier to describe and gives filters that are as good as those designed by other methods. The three steps required for the design of a length-N minimum-phase filter are as follows:

1. Design a length $= (2N - 1)$ linear-phase (type 1) filter, obtaining $A(f)$ as a sum of N cosines.
2. Scale the resulting filter by adding δ_2' so that $A(f) + \delta_2'$ is positive, where δ_2' is the stop-band error for the linear-phase filter, as shown in Fig. 4.3.
3. Factor the transfer function of the scaled filter in step 2, keeping all of the zeros that are inside the unit circle and one each of the double zeros on the unit circle.

These three steps are illustrated for a length-11 minimum-phase filter in Fig. 4.3. As described in Section 2.2.3, the linear-phase transfer function has roots with mirror-image symmetry (see Section 2.2.3). In Fig. 4.3a there are 20 roots, 14 of which are on the unit circle. The scaling in step 2 (see Fig. 4.3b) results in seven double zeros on the unit circle and does not disturb the mirror-image symmetry of the remaining six roots. Step 3 (see Fig. 4.3c) results in seven single zeros on the unit circle and three roots inside the unit circle that shape the pass band of this optimum-magnitude minimum-phase filter.

When the requirement for linear-phase is dropped along with the required symmetry of the impulse response, there may be a considerable saving in filter length for the same magnitude performance, perhaps as high as a factor of 2. The actual saving achieved depends on the specific type of filter being designed. For example, if the desired filter has a very narrow pass band, the linear-phase filter designed in step 1 will have all of its $2(N - 1)$ zeros on the unit circle. Then the

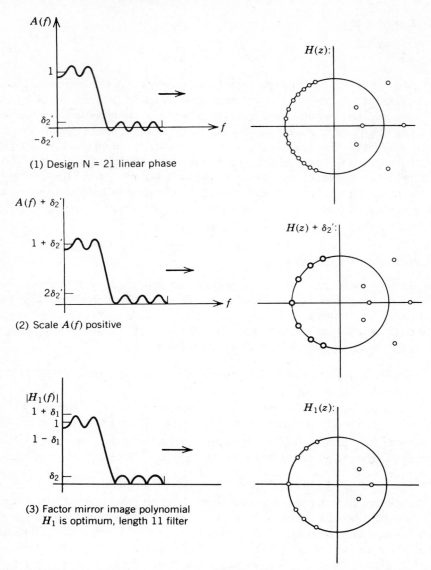

FIGURE 4.3. Design steps for optimum-magnitude, minimum-phase filter. (1) Design $N = 21$ linear phase. (2) Scale $A(f)$ positive. (3) Factor mirror image polynomial. H_1 is optimum length-11.

scaling in step 2 will result in $N - 1$ double zeros on the unit circle, so step 3 will give a filter with $N - 1$ single zeros on the unit circle. This optimum-magnitude minimum-phase filter has *linear* phase! In this special case of a very narrow pass band, the linear-phase filter is also an optimum-magnitude, minimum-phase filter with no savings as a result. Any linear-phase filter with all of its zeros on the unit circle is also an optimum-magnitude, minimum-phase filter.

The other extreme is the type of filter with a very narrow stop band (e.g., a notch filter). It is possible to reduce the length required for linear phase by more than a factor of 2, as shown in Fig. 4.4. Figure 4.4a illustrates a length-21, linear-phase notch filter with a 1.2-dB pass-band ripple and a 30-dB notch with pass-band edges at $f = 0.21$ and $f = 0.29$. A length-9, minimum-phase, optimum-magnitude filter was designed to have the same band edges, about the same pass-band ripple (1.1 dB), and a 45-dB notch, as shown in Fig. 4.4b. The group delay of the linear-phase notch filter was a constant 10 samples for all frequencies. The group delay for the minimum-phase notch was much smaller, varying between -0.5 and $+2.0$ samples.

The major difficulty in this method of minimum-phase filter design is in step 3, which requires factoring a polynomial whose order is twice the order of the desired transfer function. If we use special properties of the transfer function to be factored (e.g., that the locations of the unit circle zeros are known from the frequency domain), it is possible to design reasonably long filters (lengths greater

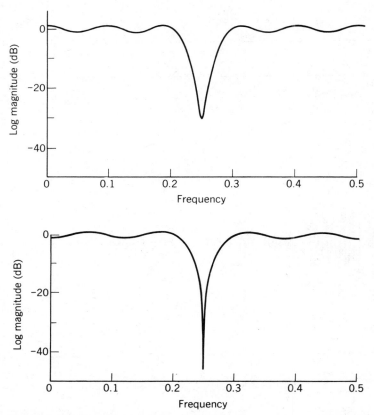

FIGURE 4.4. Notch filter comparison. (a) Linear-phase filter, $W = 21$, 1.2 dB passband ripple, 30 dB notch, passband edges 0.21 and 0.29; and (b) optimum-magnitude, minimum-phase filter, $N = 9$, 1.1 dB passband ripple, 45 dB notch, passband edges 0.21 and 0.29.

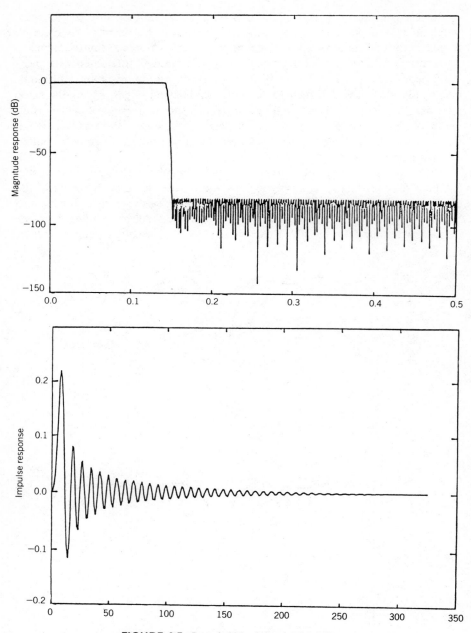

FIGURE 4.5. Length-325 minimum-phase filter.

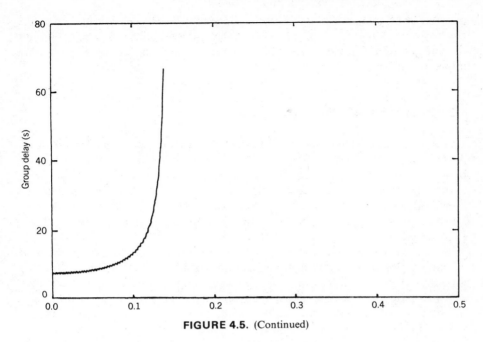

FIGURE 4.5. (Continued)

than 300).[4] For example, a length-325, minimum-phase, low-pass filter was designed with the resulting responses shown in Fig. 4.5. Figure 4.5a gives the magnitude response with about 82-dB attenuation in the stop band, and Fig. 4.5b shows the impulse response. Note that it is far from having the symmetry required for linear phase. The maximum of the impulse response occurs at sample 9, corresponding to the low-frequency group delay shown in Fig. 4.5c.

The same procedure described for obtaining an equiripple minimum-phase filter may be used to obtain a minimum-phase, Butterworth-type filter. The squared magnitude characteristic of the minimum-phase filter will have the maximally flat properties that the magnitude has for the linear-phase case. The group delay characteristic of the Butterworth minimum-phase filters is smoother than that obtained for the equiripple minimum-phase filters.

Summary

Optimum-magnitude, minimum-phase filters must have an equiripple magnitude characteristic, as shown by the minimum-phase alternation theorem. Optimum-magnitude, minimum-phase filters can be designed by factoring an appropriately scaled linear-phase prototype. Minimum-phase filters generally have smaller group delays than linear-phase filters, except for the narrow passband filters that are both linear phase and minimum phase at the same time. Minimum-phase filters generally achieve the same magnitude specifications as the linear-phase filters but with fewer coefficients. The computational saving may not be as great as it at first appears, since the linear-phase impulse response has a symmetry that allows storage of only $(N + 1)/2$ coefficients. The possibility of computational saving must be carefully examined for each particular implementation.

4.2 COMPLEX APPROXIMATION

Linear-phase and minimum-phase designs give a real approximation problem. A real-valued function of frequency f is approximated as a weighted combination of real-valued functions with real coefficients. To design filters with about the same magnitude characteristics but less delay than the linear-phase filters, we may use a complex desired function with a desired magnitude of unity and a desired group delay slightly less than that of the linear-phase filter with the same length.

When the Chebyshev error is used, the resulting approximation problem cannot be directly solved with any linear approximation scheme for real-valued functions. Steiglitz[5] has proposed a method for reformulating the complex approximation problem to allow the approximate minimization of the magnitude and phase of the error for all-pass filters. He uses linear programming for the design. This section describes a slightly different but closely related approach[6,7] that uses standard linear programming algorithms. A FORTRAN program for complex Chebyshev design is provided in the appendix (Program 7).

Least squared approximation may be used with complex-valued desired functions in exactly the same way as with the real-valued desired functions that arise in the linear-phase problem (see Section 3.2). This section describes complex LS approximation and contains an example designed with the LS complex design program in the appendix (Program 8).

4.2.1 Complex Chebyshev Error Approximation

The frequency response for a length-N FIR filter is

$$H(f) = \sum_{n=0}^{N-1} h_n e^{-jn2\pi f}, \tag{4.12}$$

where the unit-pulse response values h_n are assumed to be real. In the complex approximation problem the desired frequency response $D(f)$ is a complex-valued function of the frequency f. This leads to the

Chebyshev Complex Approximation Problem
  Given

  A compact subset $\mathscr{F}$ of $[0, 0.5]$

A desired complex-valued function $D(f)$

A positive weight function $W(f)$

the problem is to minimize over h_n

$$\max_{f \in \mathscr{F}} W(f) \left| D(f) - \sum_{n=0}^{N-1} h_n e^{-jn2\pi f} \right|, \qquad (4.13)$$

or to minimize over h_n

$$\max_{f \in \mathscr{F}} W(f) |E(f)|, \qquad (4.14)$$

where $E(f)$ is the complex-valued error function.

The main difference between this problem and the linear-phase approximation problem in Section 3.3 is that the magnitude of a complex error is to be minimized, as illustrated in Fig. 4.6, where the desired function $D(f)$ is 1.0 in the pass band and 0.0 in the stop band.

4.2.1.1. Linear Equation Approach

The complex approximation problem may be viewed as a nonlinear real approximation problem, since the minimization of the magnitude of a complex number z corresponds to the minimization of the square root of the sum of the squares of the real and imaginary parts of z; with

$$z = x + jy, \qquad |z| = \sqrt{x^2 + y^2}. \qquad (4.15)$$

The set of all points in the complex plane that have unit magnitude is a circle,

$$x^2 + y^2 = 1. \qquad (4.16)$$

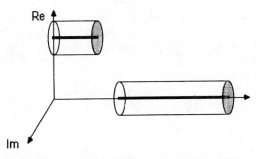

FIGURE 4.6. The complex approximation problem.

The circle implied by (4.16) may be approximated with a unit square corresponding to the equation

$$\max\{|x|, |y|\} = 1. \tag{4.17}$$

The original statement of the complex approximation problem required fitting the approximating function $H(f)$ inside the smallest, circular cross section cylinder centered on the desired function $D(f)$, as illustrated in Fig. 4.6. If a cylinder with a square cross section is used, the approximation problem in (4.14) changes, for $W(f) = 1$, to minimize over h_n

$$\max_{f \in \mathcal{F}} \{|\text{Re}\{E(f)\}|, |\text{Im}(E(f))|\}. \tag{4.18}$$

If the imaginary part of $E(f)$ is rewritten as

$$\text{Im}(E(f)) = \text{Re}(E(f)e^{-j\pi/2}), \tag{4.19}$$

equation (4.18) can be rewritten as

minimize over h_n
$$\max_{f \in \mathcal{F}} \{|\text{Re}(E(f))|, |\text{Re}(E(f)e^{-j\pi/2})|\}. \tag{4.20}$$

From this point of view the approximations of real and imaginary parts are special applications of the real rotation theorem.[6,7]

Real Rotation Theorem
For a complex number $z = x + jy$,

$$|z| = \sqrt{x^2 + y^2} = \max_{-0.5 \leqslant u \leqslant 0.5} \{\text{Re}(ze^{j2\pi u})\}.$$

The approximations of the real and imaginary parts correspond to choosing only the two values of $u = 0.0$ and $u = -0.25$ in the real rotation theorem, as shown in Fig. 4.7. Streit and Nuttall[7] have used more samples of u in the real rotation theorem to design array shading functions and have shown that 8 to 16 samples of u are sufficient for their application. Equation (4.20), when viewed as an application of the real rotation theorem, becomes

minimize over h_n
$$\max_{f \in \mathcal{F}} \max_{u \in \mathcal{U}} \{|\text{Re}(E(f)e^{-j2\pi u})|\}, \tag{4.21}$$

where the set

$$\mathcal{U} = \{0.0, 0.25\}. \tag{4.22}$$

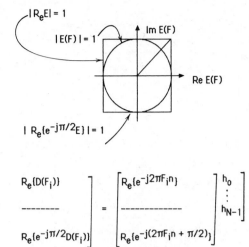

FIGURE 4.7. Approximation of real and imaginary parts.

When $E(f)$ is written out in terms of $D(f)$ and $H(f)$, (4.21) becomes

minimize over h_n

$$\max_{f \in \mathscr{F}} \max_{u \in \mathscr{U}} \left\{ \left| \text{Re}\left\{ \left(D(f) - \sum_{n=0}^{N-1} h_n e^{-j2\pi fn} \right) e^{-j2\pi u} \right\} \right| \right\}. \tag{4.23}$$

The approximation problem in (4.23) may be solved with a standard linear programming package for solving overdetermined linear equations,[8] or the improved Algorithm 635.[9] If there are L frequency values in the set $\mathscr{F}$ and two values in the set $\mathscr{U}$, as in (4.22), then there will be $2L$ equations in N unknowns, as shown in Fig. 4.7. These $2L$ equations correspond to using a square to approximate the circle as described in (4.17) and illustrated in Fig. 4.7. There are complex errors that may have a magnitude as large as 1.414 times the error minimized when the square is used to approximate the circle. As shown in Fig. 4.8, when an octagon is used to approximate the circle, the complex error magnitude is only, at worst, 1.082 times the error minimized when the circle is used. The octagonal approximation corresponds to using

$$\mathscr{U} = \{0.0, 0.125, 0.25, 0.375\} \tag{4.24}$$

in (4.21) and results in the $4L$ equations shown in Fig. 4.8.

4.2.1.2 Bandpass Design Using Complex Approximation

To illustrate the possible advantages of complex approximation, we present two bandpass designs. Program 7 in the appendix was used to design these examples.

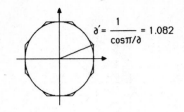

$$
\begin{bmatrix}
R_e\{D(F_i)\} \\
\text{-------} \\
R_e\{e^{-j\pi/4}D(F_i)\} \\
\text{-------} \\
R_e\{e^{-j\pi/2}D(F_i)\} \\
\text{-------} \\
R_e\{e^{-j3\pi/4}D(F_i)\}
\end{bmatrix}
=
\begin{bmatrix}
R_e\{e^{-j(2\pi F_i n)}\} \\
\text{-----------} \\
R_e\{e^{-j(2\pi F_i n + \pi/4)}\} \\
\text{-----------} \\
R_e\{e^{-j(2\pi F_i n + \pi/2)}\} \\
\text{-----------} \\
R_e\{e^{-j(2\pi F_i n + 3\pi/4)}\}
\end{bmatrix}
\begin{bmatrix}
h_o \\
\vdots \\
h_{N-1}
\end{bmatrix}
$$

4 L equations, N unknowns

FIGURE 4.8. Octagonal approximation to circle.

Example 4.2. Bandpass, Reduced Delay Filter[10]
The filter specifications are

Filter length $N = 31$
Stop-band frequencies: band 1: [0.00, 0.10]; weight = 10.0
Pass-band frequencies: band 2: [0.15, 0.28]; weight = 1.0
Stop-band frequencies: band 3: [0.33, 0.50]; weight = 10.0

The desired function $D(f)$ was

$$
D(f) = \begin{cases} e^{-j2\pi 12 f} & \text{for } f \text{ in the pass band,} \\ 0.0 & \text{for } f \text{ in the stop bands.} \end{cases}
$$

This choice of $D(f)$ corresponds to a desired group delay of 12 samples. For comparison, note that a linear-phase filter with the same length has a group delay of 15 samples. The design used a 16-sided figure to approximate the circle and took about 20 minutes of CPU time on a VAX 750, using program 7 with the standard linear programming package in reference 8. The resulting unit-pulse response is shown in Fig. 4.9. The resulting frequency response is plotted in Fig. 4.10. The pass-band deviation is $\delta_1 = 0.075$, and the stop-band deviation is $\delta_2 = 0.0075$. The pass-band group delay is between 11.08 and 13.19 samples. A

$h(1) = 0.96237558E-02$	$h(2) = 0.13125563E-01$
$h(3) = -0.23975457E-01$	$h(4) = -0.46219312E-01$
$h(5) = 0.76348322E-02$	$h(6) = 0.37956824E-01$
$h(7) = 0.26259034E-02$	$h(8) = 0.44724002E-01$
$h(9) = 0.74415697E-01$	$h(10) = -0.11377245E+00$
$h(11) = -0.23025927E+00$	$h(12) = 0.60315206E-01$
$h(13) = 0.33858679E+00$	$h(14) = 0.72350350E-01$
$h(15) = -0.25975002E+00$	$h(16) = -0.13506413E+00$
$h(17) = 0.09778991E-01$	$h(18) = 0.54102988E-01$
$h(19) = 0.73464517E-03$	$h(20) = 0.49044216E-01$
$h(21) = 0.12723715E-01$	$h(22) = -0.60300190E-01$
$h(23) = -0.26846622E-01$	$h(24) = 0.15128935E-01$
$h(25) = -0.42595769E-02$	$h(26) = 0.42152792E-02$
$h(27) = 0.80363709E-01$	$h(28) = 0.55096351E-02$
$h(29) = -0.24003784E-01$	$h(30) = -0.97490037E-02$
$h(31) = 0.47308524E-02$	

FIGURE 4.9. Unit pulse response of bandpass filter.

linear-phase filter with the same length and the same band-edge frequencies has $\delta_1 = 0.11$ and $\delta_2 = 0.011$. Thus, the complex approximation has a magnitude characteristic about 3 dB better and a group delay about three samples less than the linear-phase filter. Of course, the group delay is no longer a constant, as shown in Fig. 4.10.

It is possible to derive an approximate expression for the group delay error[10]

$$e_T(f) = \sum_{n=0}^{N-1} h_n(T_d - n)\cos(2\pi(T_d - n)f). \tag{4.25}$$

This linear expression allows additional equations to be added to the linear programming problem and to directly weight the group delay of the filter. It is also possible to weight the phase rather than the group delay.[10] The following example shows that with the use of delay weighting it is possible to get an approximately constant group delay characteristic.

Example 4.3. Bandpass, Complex Approximation with Delay Weighting
The filter specifications are

Filter length $N = 31$
Stop-band frequencies: band 1: [0.00, 0.10]; weight = 10.0
Pass-band frequencies: band 2: [0.15, 0.28]; weight = 1.0
Stop-band frequencies: band 3: [0.33, 0.50]; weight = 10.0
The weight on the group delay was 1.0
The desired function $D(f)$ was

$$D(f) = \begin{cases} e^{-j2\pi 12f} & \text{for } f \text{ in the pass band,} \\ 0.0 & \text{for } f \text{ in the stop bands.} \end{cases}$$

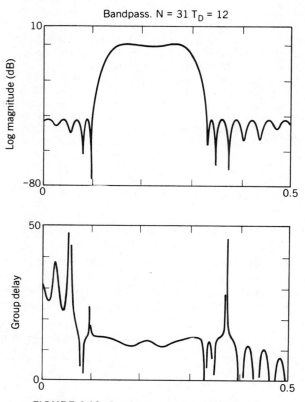

FIGURE 4.10. Bandpass complex approximation.

These specifications were used in Example 4.2, but here there is an additional weighting on the group delay from (4.25).

The resulting pass-band and stop-band errors were 0.11 and 0.011, which are the same as those obtained with a linear-phase filter. The group delay in the pass band was between 11.90 and 12.13, as shown in Fig. 4.11. This filter has about the same magnitude as the linear-phase filter, but the approximately constant group delay is only 12 instead of 15. The unit pulse-response of the filter is shown in Figure 4.12.

4.2.1.3 Equalizer Design Using Complex Approximation

When an FIR filter is used to equalize or compensate an existing system or filter, the problem is usually a complex approximation problem. A system with good magnitude characteristics but bad group delay characteristics may be followed by a FIR equalizer that will add a group delay characteristic that can make the overall equalized delay close to the desired characteristic. The possibilities are illustrated by an equalized fourth-order elliptic filter in Example 4.4.

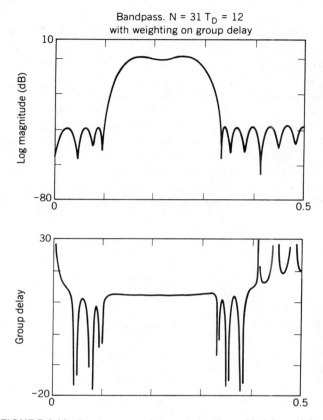

FIGURE 4.11. Bandpass complex approximation with delay weight.

$h(\ 1) = \ \ 0.12270841E-01$	$h(\ 2) = \ \ 0.21234122E-01$
$h(\ 3) = -0.30222217E-01$	$h(\ 4) = -0.56719187E-01$
$h(\ 5) = \ \ 0.10123622E-01$	$h(\ 6) = \ \ 0.42773458E-01$
$h(\ 7) = \ \ 0.16342754E-02$	$h(\ 8) = \ \ 0.47894601E-01$
$h(\ 9) = \ \ 0.81674523E-01$	$h(10) = -0.12347055E+00$
$h(11) = -0.24866632E+00$	$h(12) = \ \ 0.65881834E-01$
$h(13) = \ \ 0.33825492E+00$	$h(14) = \ \ 0.74176101E-01$
$h(15) = -0.24832359E+00$	$h(16) = -0.12874673E+00$
$h(17) = \ \ 0.31974335E-01$	$h(18) = \ \ 0.49055367E-01$
$h(19) = \ \ 0.77787742E-03$	$h(20) = \ \ 0.45204235E-01$
$h(21) = \ \ 0.12740520E-01$	$h(22) = -0.56494463E-01$
$h(23) = -0.28209696E-01$	$h(24) = \ \ 0.20944360E-01$
$h(25) = \ \ 0.12019594E-01$	$h(26) = \ \ 0.26787085E-03$
$h(27) = \ \ 0.52933804E-02$	$h(28) = \ \ 0.24858904E-02$
$h(29) = -0.39502040E-02$	$h(30) = \ \ 0.20514977E-02$
$h(31) = -0.21685219E-02$	

FIGURE 4.12. Unit pulse response for delay weighting.

Example 4.4. Equalization of a Fourth-Order Elliptic Filter

The fourth-order elliptic filter shown in Fig. 4.13 has a pass-band edge of $f_p = 0.25$ and a stop-band edge of $f_s = 0.3$ with a pass-band deviation of 0.5 dB and a stop-band attenuation of at least 34 dB. This filter is minimum phase with a group delay that varies by about 11 samples in the pass band. As shown in Fig. 4.14, the group delay increases rapidly near the band edge.

A length-31 FIR equalizer was designed with the complex approximation algorithm to obtain an equalized delay of 12.9 samples with a delay error of only 0.65 samples in the pass band. In addition to equalizing the delay, the FIR equalizer also provides additional attenuation in the stop band of 15.6 dB, since the pass-band and stop-band deviations of the equalizer are 0.16. The equalizer magnitude is shown in Fig. 4.15. The group delay of the elliptic filter before and after equalization is shown along with the group delay of the equalizer in Fig. 4.16.

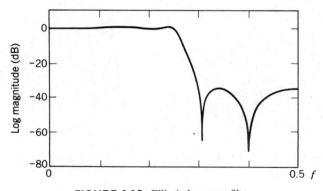

FIGURE 4.13. Elliptic low-pass filter.

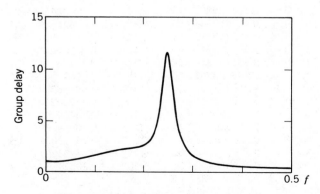

FIGURE 4.14. Group delay of elliptic filter.

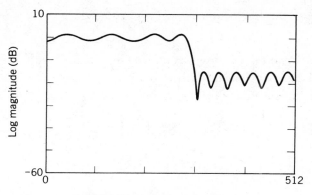

FIGURE 4.15. Equalizer magnitude.

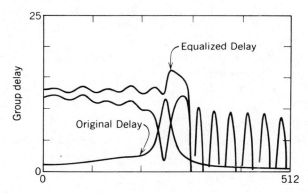

FIGURE 4.16. Original and equalized delay together with the equalizer delay.

4.2.2 Complex Approximation with Least Squared Error

The LS approximation theory described in Section 3.2.1 for linear-phase filter design can also be used for design of filters with arbitrary-phase characteristics.

The frequency response for a length-N FIR filter is

$$H(f) = \sum_{n=0}^{N-1} h_n e^{-jn2\pi f}, \tag{4.26}$$

where we assume that the unit-pulse response values h_n are real. In the complex approximation problem the desired frequency response $D(f)$ is a complex-valued function of the frequency f. As in Section 3.2.1, the error is minimized over a set of L discrete frequencies f_k.

Least Squared Complex Approximation Problem
Given

A set of discrete frequencies f_k contained in $[-0.5, 0.5]$
A desired complex-valued function $D(f_k)$
A positive weight function $W(f_k)$

the problem is to minimize, by choice of h_n,

$$\sum_{k=0}^{L-1} W(f_k) \left| D(f_k) - \sum_{n=0}^{N-1} h_n e^{-jn2\pi f_k} \right|^2. \tag{4.27}$$

We can obtain LS solution to the overdetermined set of linear equations in (4.27) by using standard techniques.[11] Program 8 solves the LS problem by using a subroutine from LINPACK.[11] This program was used to design a bandpass filter with the same band edges, weights, and desired delay as Example 4.2.

Example 4.5. Bandpass, Reduced Delay Filter, Least Squared Error
The filter specifications are

Filter length $N = 31$
Stop-band frequencies: band 1: $[0.00, 0.10]$; weight $= 10.0$
Pass-band frequencies: band 2: $[0.15, 0.28]$; weight $= 1.0$
Stop-band frequencies: band 3: $[0.33, 0.50]$; weight $= 10.0$
The desired function $D(f)$ was

$$D(f) = \begin{cases} e^{-j2\pi 12 f} & \text{for } f \text{ in the pass band,} \\ 0.0 & \text{for } f \text{ in the stop bands.} \end{cases}$$

This choice of $D(f)$ corresponds to a desired group delay of 12 samples. For comparison, note that a linear-phase filter with the same length has a group delay of 15 samples. The resulting frequency response is plotted in Fig. 4.17. The use of transition bands has greatly reduced the error peaks at the band edges, as shown in the magnitude plot of Fig. 4.17. The group delay varied between 11.47 samples and 13.19 samples, an error very similar to that obtained in Example 4.2. The LS design took only about one tenth of the computing time taken by the Chebyshev design in Example 4.2.

Summary

By complex approximation theory we can design filters that have smaller group delays than linear-phase filters and less delay distortion than minimum-phase

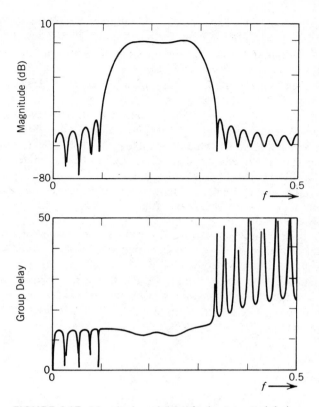

FIGURE 4.17. Magnitude and delay for least-squared design.

filters. Complex approximation techniques can also be used to equalize magnitude and phase characteristics of a given system. The complex approximation problem can be solved by the Chebyshev or the LS error criterion.

The Chebyshev approximation problem was changed into a problem of finding the approximate solution to an overdetermined set of linear equations. This problem was in turn solved by standard linear programming techniques.

The LS approach is often appropriate when performance is measured in terms of signal energy. Complex LS designs are done with Program 8, which uses the same LINPACK subroutines used in Chapter 3 for linear-phase design.

REFERENCES

[1]   O. Herrmann and H. W. Schüssler, "Design of Nonrecursive Digital Filters with Minimum-Phase," *Electronic Lett.* **6**, 329–330 (1970).

[2]   R. Boite and H. Leich, "A New Procedure for the Design of High Order Minimum-Phase FIR Digital or CCD Filters," *Signal Processing* **3**, 101–108 (1981).

[3] G. A. Mian and A. P. Nainer, "A Fast Procedure to Design Equiripple Minimum-phase FIR Filters," *IEEE Trans. Circuits Systems* **29**, 327–331 (1982).

[4] X. Chen and T. W. Parks, "Design of Optimal Minimum Phase FIR Filteers," *Signal Processing* Vol. 10, pp. 369–383 June 1986.

[5] K. Steiglitz, "Allpass FIR Phase Equalizers," *IEEE Trans. Acoust., Speech, Signal Processing* 125–129 (1982).

[6] K. Glashoff and K. Roleff, "A New Method for Chebyshev Approximation of Complex-Valued Functions," *Math. Comp.* **36**, 233–239 (1981).

[7] R. L. Streit and A. H. Nuttall, "A General Chebyshev Complex Function Approximation Procedure and an Application to Beamforming," *J. Acoust. Soc. Am.* **72**, 181–190 (1982).

[8] J. Barrodale and C. Phillips, "Solution of an Overdetermined System of Linear Equations in the Chebyshev Norm, Algorithm 495," *ACM Trans. Math. Software* **1**, 264–270 (1975).

[9] R. L. Streit, "An Algorithm for the Solution of Systems of Complex Linear Equations in the L_∞ Norm with Constraints on the Unknowns," Algorithm 635, ACM Trans. Math. Software, **11**, 242–249 (1985).

[10] X. Chen and T. W. Parks, "Design of FIR Filters in the Complex Domain," *IEEE Trans. Acoust., Speech, Signal Processing*, **35**, 144–153 (1987).

[11] J. J. Dongarra, J. R. Bunch, C. B. Moler, and G. W. Stewart, *LINPACK User's Guide*, Philadelphia: SIAM, 1979.

5
Implementation of Finite Impulse-Response Filters

This chapter contains three major sections. The first section discusses representation of continuous-amplitude signal samples in terms of discrete-amplitude or digital samples. The next section describes various ways of implementing an FIR digital filter. These implementations are given in terms of difference equations, block diagrams (structures), and assembly language programs. The final section treats the finite word-length effects of coefficient quantization and quantization noise.

When a filter is implemented with a digital computer or digital hardware, the signal and coefficient values can no longer be represented with arbitrary precision and unlimited amplitude. Numbers must be represented as members of a finite set of values in a digital processor. There are several schemes for approximately representing real numbers digitally, but principally floating-point and fixed-point representations are used. The minimum computing time, or the most powerful filter that can be computed in a given time, is usually best obtained by fixed-point arithmetic. Furthermore, most signal-processing chips use fixed-point arithmetic to efficiently use the limited silicon area available. This book analyzes fixed-point implementations of digital filters. More complete treatment of finite word-length effects can be found in the recently published texts 1 and 2 and in reference 3. The presentations in this chapter and in Chapter 8 have been motivated by the work of H. W. Schüssler and by reference 4.

Finite word-length effects may be divided into two different categories[1,2]:

1. Errors in representing coefficients as finite fixed-point numbers. (The actual filter does not have exactly the correct coefficients but is still linear.)
2. Errors due to the finite-precision arithmetic operations of addition, multiplication, and storage. (These errors make the digital filter a nonlinear system.)

These two types of finite word-length errors require very different analysis techniques. For the error in 1, linear analysis can be used; but for the errors in 2, nonlinear analysis methods must be used.

The discussion of quantization errors begins with an analysis of the conversion of an analog voltage with continuous-amplitude values to a digital representation with discrete values. The discussion on digital filter structures relates equations for computing the filter output to structures and programs for implementing the filter on a programmable signal processor. Nonrecursive filters do not generally have severe quantization problems; therefore, the discussion of nonrecursive implementations in this chapter is brief. Recursive filters, however, have problems with coefficient sensitivity, quantization noise, and quantization-induced instabilities. Chapter 8 treats these possible problems with emphasis on second-order blocks.

5.1 DIGITAL SIGNAL REPRESENTATIONS

Digital filtering requires that signals be both discrete time and discrete amplitude. The conversion from continuous time to discrete time is called *sampling*. The discrete-time signal produced by sampling is then converted to a discrete-amplitude signal by a process called *analog-to-digital* (A/D) *conversion*. Analog-to-digital conversion, as described here, has nothing to do with the time variable. A sample value, a real number that may take on a nondenumerably infinite number of values, is approximated by or "converted to" a digital number that can only take on one of a finite set of values.

5.1.1. Two's Complement Arithmetic

In the basic binary number representation of the integer x,

$$x = \sum_{n=0}^{B-1} b_n 2^n, \tag{5.1}$$

the bits $b_m, m = 0, \ldots, B - 1$, are either 1 or 0—hence the name *binary*. The bits are written $b_{B-1} \cdots b_0$, where the leftmost bit b_{B-1} is called the *highest-order bit* or the *most significant bit*.

To map the infinite range of values of the real number x into a finite range, we evaluate the value of x modulo 2^B. Two's complement arithmetic is really arithmetic modulo 2^B. Any number outside of the range of $1, \ldots, 2^{B-1}$ is reduced to this range by subtracting an appropriate integer multiple of 2^B. Intermediate results in a computation may overflow, and the correct output will still be obtained, provided that the output is within the range of $1, \ldots, 2^{B-1}$.

Negative numbers are represented as the additive inverses of the positive numbers. For example, when 1 is added to $2^B - 1$, the result is 2^B, which is equivalent to zero modulo 2^B. Thus, $2^B - 1$ is identified as "-1." In B-bit two's complement arithmetic, the numbers up to and including $2^{B-1} - 1$ represent positive numbers.[1,2] They all have a highest-order bit of 0. The next number, 2^{B-1}, is the most negative number in the two's complement system. The numbers from 2^{B-1} up to and including $2^B - 1$ all have a highest-order bit of 1 and represent negative numbers. The circle of 3-bit two's complement numbers in Fig. 5.1a shows modulo-8 arithmetic. The integer representations for the binary numbers are shown on the inside of the circle.

Figure 5.1a shows that as x increases, the representation wraps around the circle. Information is lost about the number of times that x has wrapped around the circle. Only the 3-bit residue of x modulo 8, the relative position on the circle, is available.

5.1.2. Fractions

For easy truncation or rounding, the usual way to describe and use fixed-point arithmetic is with fractions. If the largest 3-bit positive number is thought of as $\frac{3}{4}$, and the most negative number as $\frac{4}{4} = -1.000$, then the product of any two numbers is a number whose magnitude is less than or equal to 1. The highest-order bit is called the *sign bit*. A real-valued voltage v, between $-V$ and $+V$ volts, is represented in two's complement arithmetic by B bits with a fractional part

$$V(2^{-(B-1)})\left(\sum_{n=0}^{B-2} b_n 2^n\right) \tag{5.2}$$

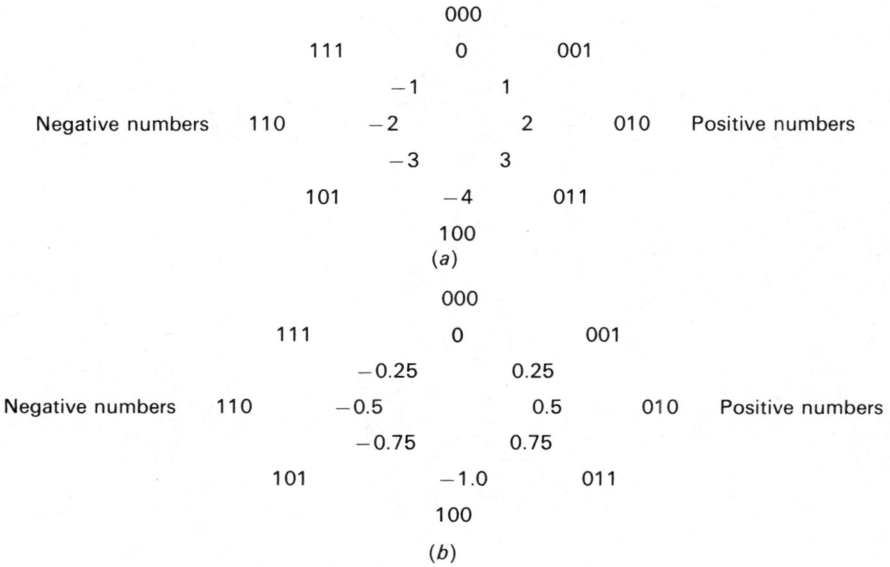

FIGURE 5.1. Circles of 3-bit two's complement numbers. (a) Integers and (b) fractions.

as the quantized value

$$[v]_Q = -Vb_{B-1} + V \sum_{n=0}^{B-2} b_n 2^{-B+1+n}, \qquad (5.3)$$

where each of the B bits, b_n, $n = 0, \dots, B - 1$, is either 1 or 0. Figure 5.1b shows the fractional representations for 3-bit numbers. For example, the bit pattern 101 in Fig. 5.1b with $b_0 = 1$, $b_1 = 0$, and $b_2 = 1$ represents the number

$$-1 + 2^{-2} + 0 = -0.75.$$

5.1.3 Quantization Error

The approximate representation of the real number v in (5.3), $[v]_Q$, must be one of the 2^B-possible values of the fraction in (5.3). The separation between adjacent quantized values, known as the *quantization step size*, is

$$Q = V(2^{-(B-1)}). \qquad (5.4)$$

There is a nonlinear relation between v and $[v]_Q$ that depends on whether the approximation to v is made by truncation or rounding. The relation between the voltage v and its quantized approximation is shown in Fig. 5.2 for a 3-bit

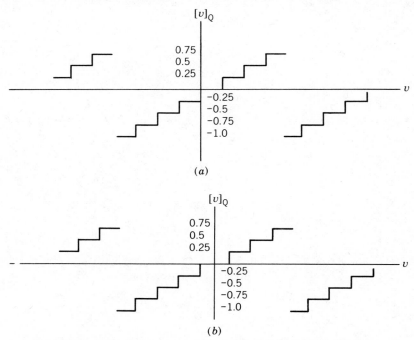

FIGURE 5.2. Quantization with three bits. (*a*) Truncation and (*b*) rounding.

representation with $V = 1$, that uses both truncation (5.2a) and rounding (5.2b). The maximum value of $[v]_Q$ is $1 - Q = 0.75$, and the minimum value is -1.0.

The periodic nature of the two's complement type of overflow behavior is also illustrated in these figures. After the voltage v exceeds $+0.75$ V, it is represented as -1 V (see Fig. 5.2a). The periodicity shown in Fig. 5.2 corresponds to the wraparound described in connection with Fig. 5.1.

Quantization of a signal is a memoryless nonlinear operation. The input to the memoryless nonlinear system, shown in Fig. 5.3, is the signal voltage v, and the output is the quantized signal $[v]_Q$. Although the quantization process is deterministic, the difference between v and $[v]_Q$ is usually modeled as a random variable

$$n = [v]_Q - v. \tag{5.5}$$

The quantized signal is considered to be the true signal v with an added noise component n, as shown in Fig. 5.3b. This quantization noise can be modeled as a uniformly distributed random variable that is independent of the signal v when the number of bits is reasonably large, the error is relatively small, and the signal is changing rapidly enough from sample to sample.[2,3]

For truncation, the quantization error n lies between 0 and Q and is modeled as a uniformly distributed random variable with a mean value of $Q/2$. For rounding, the quantization error or noise is modeled as a uniformly distributed random variable with zero mean. The assumed probability densities for truncation and rounding are shown in Fig. 5.4.

The variance of the random variable n is given by

$$\sigma_n^2 = E\{n - E\{n\}\}^2, \tag{5.6}$$

where $E\{x\}$ is the expected value of x. For rounding, the noise has zero mean, $E\{n\} = 0$, and the variance

$$\sigma_n^2 = \int p(u)u^2 \, du. \tag{5.7}$$

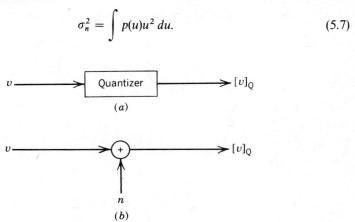

FIGURE 5.3. Modeling quantization noise. (a) Nonlinear; (b) linear model.

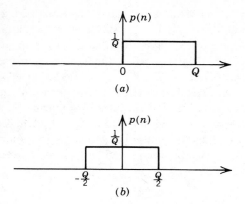

FIGURE 5.4. Probability densities (*a*) Truncation and (*b*) rounding.

Using the probability density $p(u)$, shown in Fig. 5.4*b*, gives

$$\sigma^2 = \frac{1}{Q} \int_{-Q/2}^{Q/2} u^2 \, du = \frac{Q^2}{12}. \tag{5.8}$$

The variance for truncation is also $Q^2/12$.

The errors that are made in converting a continuous-amplitude signal into a discrete representation may be evaluated in terms of a signal-to-noise ratio (SNR). The signal must be scaled to limit the possibility of overflow with the use of a scale factor or gain factor G, as shown in Fig. 5.5. A small value of G will ensure that overflow never occurs, but the SNR will be reduced because the quantization noise level is fixed and a small value of G reduces the signal component. If occasional overflow is allowed, then the signal component will be larger and thus the SNR will be increased. This tradeoff between overflow and quantization noise is always necessary when using fixed-point arithmetic.

With $V = 1$ in (5.3) the quantization step size, from (5.4), is

$$Q = 2^{-B+1}. \tag{5.9}$$

The noise variance is

$$\sigma_n^2 = \frac{Q^2}{12} = \frac{2^{-2B}}{3}. \tag{5.10}$$

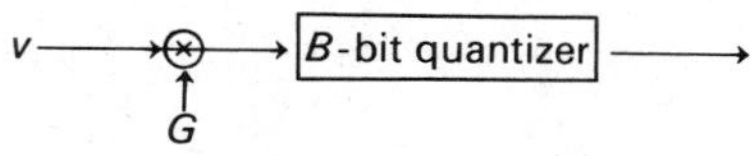

$$v \longrightarrow \otimes \longrightarrow \boxed{B\text{-bit quantizer}} \longrightarrow$$
$$\uparrow$$
$$G$$

FIGURE 5.5. Signal scaling.

The SNR is defined as

$$SNR = 10 \log \left[\frac{E\{(Gv)^2\}}{\sigma_n^2} \right].$$ (5.11)

Using the values in (5.10), we obtain

$$SNR = 10 \log[E\{(Gv)^2\}] - 10 \log \left[\frac{2^{-2B}}{3} \right]$$ (5.12)

$$= 10 \log[E\{(Gv)^2\}] + 20B \log 2 + 4.77.$$ (5.13)

The SNR clearly depends on the signal statistics. A reasonable assumption, based on the central limit theorem (CLT), is that the signal is a Gaussian random variable[5] with mean zero and variance σ^2. If G is chosen to be

$$G = \frac{1}{4\sigma},$$ (5.14)

overload will only occur 64 times in a million samples according to the Gaussian probability law. In other words, the probability that a Gaussian random variable falls within the 4σ range[5] is 0.999936. Substituting (5.14) in (5.13) gives

$$SNR = 10 \log[\tfrac{1}{16}] + 6.02B + 4.77,$$ (5.15)

which is approximately

$$SNR \cong 6B - 7.3 \text{ dB}.$$ (5.16)

The exact value of SNR depends on the choice of G. A larger value of G would give a larger value for SNR but would increase the probability of overflow. Conversely, if G were reduced to a value smaller than in (5.14), the probability of overflow would be reduced, but the SNR would also be reduced. If the signal samples were governed by a different probability law, slightly different results would be obtained. A good rule of thumb is to assume the SNR to be about 6 dB/bit.

Summary

This section introduced the concepts of two's complement arithmetic, fractional representation of numbers, and quantization noise. The tradeoff between scaling and quantization noise was discussed. The signal should be scaled to be as large as possible consistent with the allowed frequency of overflow. In this way all available quantization levels are used, and the ratio of signal to quantization noise is maximized.

5.2 EQUATIONS, STRUCTURES, AND PROGRAMS

After an FIR filter has been designed by the techniques in Chapters 3 and 4, the approximation problem has been solved. The coefficients in the filter transfer function have been calculated to meet a given specification. The second part of digital filter design is the realization problem. The transfer function of the filter must be "realized" as a piece of digital hardware or as a program to implement the input/output relation implied by the filter transfer function.

For a given transfer function there are many different ways to implement or program the digital filter. These various implementations are represented with block diagrams and are called *filter structures*. This section describes two different structures for FIR digital filters and relates these structures to assembly language programs to implement the filter.

Many different factors enter into the selection of a particular structure for a particular application. One structure may be preferred over the other because it is easier to program for a particular computer or signal-processing chip. The choice of structure may be made according to the regularity of the VLSI implementation. One structure may be less sensitive to errors in coefficients. A structure may be chosen to minimize noise introduced by quantization of the signal.[1]

This section relates the filter transfer function, the equations for calculating the output from the input, block diagrams, and programs for a digital signal-processing chip. One of the simplest digital filters is the length-3 FIR filter with a transfer function

$$H(z) = h_0 + h_1 z^{-1} + h_2 z^{-2}. \tag{5.17}$$

The equation that provides the output is the convolution

$$y(n) = h_0 x(n) + h_1 x(n-1) + h_2 x(n-2). \tag{5.18}$$

The structure indicated by the block diagram in Fig. 5.6 illustrates the direct calculation of (5.18) and is called the *direct structure*.

The boxes labeled z^{-1} in Fig. 5.6 represent unit sample delays. The value $x(n-1)$ is a delayed version of $x(n)$. If the filter were implemented with a tapped delay line, the z^{-1} would correspond to a physical delay element. However, when a digital computer program is written to implement (5.18), the boxes

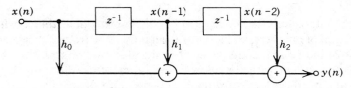

FIGURE 5.6. Direct nonrecursive structure.

```
NEXT   IN     XN,ADC    Read input x(n) from A/D converter
       LT     XN2       Load temporary register with x(n-2)
       MPY    H2        Multiply x(n-2) by h2
       PAC              Load h2x(n-2) into accumulator
       LT     XN1       Load temporary register with x(n-1)
       DMOV   XN1       x(n-1) moved to location XN2
       MPY    H1        Multiply x(n-1) by h1
       APAC             h1x(n-1) added to accumulator
       LT     XN        Load temporary register with x(n)
       DMOV   XN        x(n) moved to location XN1
       MPY    H0        Multiply x(n) by h0
       APAC             h0x(n) added to accumulator
       SACH   YN,1      Store contents of accumulator in YN
       OUT    YN,DAC    Put out y(n) to D/A converter
       CALL   WAIT      Wait for next input
       B      NEXT      Go back and get next input
```

FIGURE 5.7. Assembly code for direct structure.

labeled z^{-1} correspond to storage of variables rather than any delay. This is illustrated in Fig. 5.7, where assembly language instructions for the TMS32010 signal-processing chip are shown. (See reference 6 for a detailed description of the instructions.) The program assumes that the present input $x(n) = XN$ and the two most recent inputs $x(n-1) = XN1$ and $x(n-2) = XN2$ are stored in memory.

The blocks in Fig. 5.6 labeled z^{-1} correspond to the DMOV XN1 and DMOV XN instructions, which shift the data after it has been used. The code in Fig. 5.7 is presented to explain how to implement the direct structure. A shorter (and faster) program can be written by using a special instruction (LTD), which performs the operations of the three instructions APAC, LT, and DMOV. The LTD instruction is used in the design example at the end of this chapter.

Another structure, called the *transpose structure* because the matrix form is the transpose of that in Fig. 5.7,[1] implements exactly the same input/output relation (5.18). A block diagram of the transpose structure is shown in Fig. 5.8. The structure in Fig. 5.8 leads to a very different program for computing the filter output. Assembly code corresponding to the transpose structure is shown for the TMS32010 in Fig. 5.9. The delay blocks in Fig. 5.8 correspond to the instructions SACH Z1 and SACH Z2 in Fig. 5.9. The programs for the direct and transpose structures each have the same number of instructions and take the same amount of time to run. However, the direct structure can better take

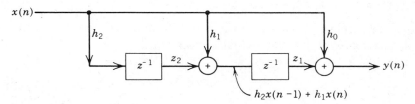

FIGURE 5.8. Transpose nonrecursive structure.

```
NEXT    IN    XN,ADC      Read x(n) from A/D converter
        LT    XN          Load temporary register with x(n)
        MPY   H0          Multiply x(n) by h0
        LAC   Z1,15       Load z1 into high accumulator
        APAC              (h0x(n) + z1) now in the accumulator
        SACH  YN,1        (h0x(n) + z1) stored in YN
        MPY   H1          Multiply x(n) by h1
        LAC   Z2,15       Load z2 into high accumulator
        APAC              (h1x(n) + z2) now in the accumulator
        SACH  Z1,1        (h1x(n) + z2) stored in Z1
        MPY   H2          Multiply x(n) by h2
        PAC               h2x(n) now in the accumulator
        SACH  Z2,1        h2x(n) stored in Z2
        OUT   YN,DAC      Output to D/A converter
        CALL  WAIT        Wait for next input
        B     NEXT        Go back and get next input
```

FIGURE 5.9. Assembly code for transpose structure.

advantage of the special LTD instruction and has less quantization noise; it is thus preferred for FIR filtering.

Summary

The way that a digital filter computes its output can be described with difference equations, block diagrams describing a structure for a discrete-time system, or computer programs. This section related these three representations and described the direct and transpose nonrecursive structures for FIR filters.

5.3 FINITE WORD-LENGTH EFFECTS IN FILTER IMPLEMENTATION

The direct and transpose structures described in Section 5.2 are nonrecursive implementations of an FIR filter. This section discusses the two categories of finite word-length effects in nonrecursive filters. First, the errors introduced by quantization of the filter coefficients are analyzed as the addition of an error system to the ideal system with unquantized coefficients. Scaling to avoid overflow is then discussed and related to the problem of maximizing the ratio of signal to quantization noise at the output of the filter.

5.3.1 Coefficient Quantization

The coefficients in the nonrecursive filter must be quantized to B_1 bits. Instead of implementing (5.18), the filter actually implements

$$y(n) = \sum_{m=0}^{N-1} [h(m)]_Q x(n - m), \tag{5.19}$$

where $[h(m)]_Q$ represents quantized filter coefficients.

The frequency response with quantized coefficients $\tilde{H}(f)$ may be viewed as the sum of the ideal (unquantized) response and the frequency response of an error system $H_e(f)$.

$$\tilde{H}(f) = \sum_{m=0}^{N-1} h(m)e^{-j2\pi fm} + \sum_{m=0}^{N-1} \{[h(m)]_Q - h(m)\}e^{-j2\pi fm} \qquad (5.20)$$

$$= H(f) + H_e(f). \qquad (5.21)$$

The maximum value of the response of the error system is bounded by the inequality

$$|H_e(f)| \leqslant N \max_m |\{[h(m)]_Q - h(m)\}|. \qquad (5.22)$$

When the coefficients are rounded to B_1 bits,

$$|H_e(f)| \leqslant N2^{-B_1}. \qquad (5.23)$$

The addition of the error system may limit the attenuation in the stop band, for example. In other words, the error system may allow additional signal transmission in the desired stop band. From (5.23) we find the maximum possible stop-band transmission, in dB, to be bounded by

$$20 \log_{10}|H_e(f)| \leqslant 20 \log_{10}(N2^{-B_1}). \qquad (5.24)$$

Since $20 \log_{10}(2) \cong 6$, this bound simplifies to

$$20 \log_{10}N + 20 \log_{10}(2^{-B_1}) = 20 \log_{10}N - 6B_1 \text{ dB}, \qquad (5.25)$$

giving the bound

$$20 \log_{10}|H_e(f)| \leqslant 20 \log_{10}N - 6B_1 \text{ dB}, \qquad (5.26)$$

where B_1 is the number of bits used to represent the filter coefficients in the length-N filter. For example, with 16-bit coefficients in a length-100 filter ($N = 100$, $B_1 = 16$), (5.26) shows that $H_e(f)$ may be as large as -56 dB.

The bound in (5.26) is very conservative, and in most cases one can get by with fewer bits if an optimization procedure is used to pick the best *quantized* coefficients rather than simply rounding the coefficients determined from a program that assumes no coefficient errors.[7,8] Generally, with 16-bit coefficients and short filters, the rounded coefficients will be adequate. However, when very few bits are used for filter coefficients, then we can obtain significant improvement over the rounded values by using an optimization program.[7]

5.3.2 Scaling and Overflow

The direct implementation of a length-3 nonrecursive filter is shown in Fig. 5.10. This figure has the same structure as Fig. 5.6; however, it has been redrawn to emphasize the single output accumulator. The output of the filter at time n is given by

$$y(n) = \sum_{m=0}^{N-1} h(m)x(n-m). \tag{5.27}$$

When the input $x(n)$ and the unit-pulse response $h(n)$ have magnitudes less than or equal to unity, the magnitude of $y(n)$ in (5.27) is bounded by

$$|y(n)| \leq \sum_{m=0}^{N-1} |h(m)|\,|x(n-m)|. \tag{5.28}$$

For an input with magnitude at most unity, the largest possible value of the output at time n occurs when

$$x(n-m) = \text{sgn}[h(m)]. \tag{5.29}$$

In the worst case

$$y(n) = \sum_{m=0}^{N-1} |h(m)|. \tag{5.30}$$

This equation is known as the l_1 norm of h:

$$\|h\|_1 = \sum_n |h(n)|. \tag{5.31}$$

If the unit-pulse response samples are all divided by the scale factor

$$G = \|h\|_1 \tag{5.32}$$

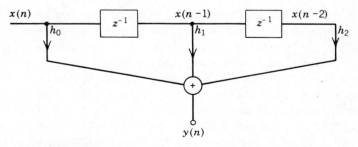

FIGURE 5.10. Direct implementation of a nonrecursive filter.

to give the scaled unit-pulse response

$$\tilde{h}(n) = \frac{h(n)}{G}, \tag{5.33}$$

Then the maximum output magnitude of the scaled filter will be less than or equal to unity, and overflow will be completely avoided.

This worst-case bound is slightly conservative for two reasons. First, since the largest positive signal value is $1 - Q$, where Q is the quantization step size, (5.30) cannot quite be attained unless all of the filter coefficients are negative. Second, it is very unlikely that the worst-case signal, (5.29), will ever occur in practice.

We can calculate the gain factor by using one of the following two additional measures of the size of $h(n)$ to give a less-conservative scaling rule. Both of these measures of gain are based on norms of the unit-pulse response h.

The l_2 norm of h,

$$\|h\|_2 = \left[\sum_n h^2(n) \right]^{1/2}, \tag{5.34}$$

is always less than or equal to the l_1 norm of h. The Chebyshev norm of the frequency response $H(f)$,

$$\|H\|_C = \max_f |H(f)|, \tag{5.35}$$

is also always less than the l_1 norm of h.

If $G = \|h\|_1$, then the signal at the output of Fig. 5.10 is guaranteed not to overflow. Since the l_2 norm of h is less than or equal to the l_1 norm of h when $G = \|h\|_2$ is used, larger unit-pulse response values result and the output SNR is improved. This improved SNR comes at the expense of the possibility of overflow. The choice of gain $G = \|H\|_C$ only guarantees that the steady-state response of the system to a sine wave will not overflow. Transient signals may occasionally cause overflows. However, the frequency-domain scaling measure is easier to interpret than the other two norms and is often the preferred method for calculating scaling factors.[4] The scaling procedure is described in detail in the design example for a length-21 filter implemented with the direct structure.

5.3.3 Quantization Noise

In Fig. 5.10 it is good practice to accumulate the sum in double precision, reducing to the original word length only for additional processing or storage of the output $y(n)$. If a double-precision accumulation is not performed, quantization errors will be introduced when the low-order bits are discarded. The quantized signal $y_Q(n)$ is an approximation to $y(n)$ without signal quantization.

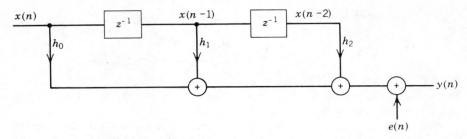

FIGURE 5.11. Direct implementation of a nonrecursive filter with quantization noise.

In Fig. 5.11 the quantization error or noise

$$e(n) = y_Q(n) - y(n) \tag{5.36}$$

is shown added to the correct output $y(n)$. If single precision were used, there would be two sources of quantization noise.

This quantization noise can be modeled as uniformly distributed, independent random variables that are independent of the signal $y(n)$ when the number of bits is reasonably large, the error is relatively small, and the signal is changing rapidly enough from sample to sample.[1,2] This error can be analyzed in the same way as the error in Section 5.1.3. It is easy to understand the effect of the quantization error in Fig. 5.11, because it occurs at the output of the filter and is represented as an external white-noise source with variance

$$\sigma_n^2 = \frac{2^{-2B}}{3}. \tag{5.37}$$

However, when the filter is implemented in the transpose structure of Fig. 5.8, quantization noise is introduced inside the filter with the variable z_1. For the transpose structure there are two sources of quantization noise, as shown in Fig. 5.12. Because of the extra source of quantization noise in the transpose structure, the direct form is recommended for nonrecursive filters.

There are many other structures for FIR filters. For example, the transfer function can be factored, and the resulting shorter filter sections can be

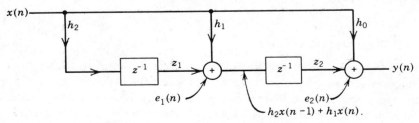

FIGURE 5.12. Quantization noise in the transpose structure.

cascaded.[4] When the filter coefficients have certain symmetries, as do linear-phase filters, special structures may be appropriate. However, for the TMS320 signal-processing chip and for most signal processors with single-cycle multiply capability and adequate word lengths, the direct structure is usually the best choice.

Summary

Coefficient quantization was analyzed in terms of an error system added to the ideal system without coefficient quantization. It was shown that the magnitude of the frequency response of the error system has an upper bound that increases with increases in filter length and decreases when more bits are used to represent the filter coefficients.

Scaling strategies based on the l_1 norm of the unit-pulse response (most conservative), the l_2 norm of the unit-pulse response, and the Chebyshev norm of the frequency response were described for reducing or eliminating overflow.

Quantization noise in the output of an FIR direct structure or a transpose structure can be analyzed in the same way as quantization noise is analyzed in Section 5.1.

5.4 DESIGN EXAMPLE

This design example presents the step-by-step procedure for designing a length-21 low-pass filter. First, one enters the specifications into the design program (Program 6) provided in the appendix. After the coefficients are calculated, scaling is performed to prevent overflow. Finally, one writes the TMS32010 assembly language program to implement the filter, using the direct structure shown in Fig. 5.6.

STEP 1. The first step in the design is to decide on the filter specifications. For this example the specifications are those of Example 3.14, the length-21 low-pass filter. The specifications and the output of Program 6 are repeated here in Fig. 5.13 for convenience.

STEP 2. The next step is to decide on the structure to be used in implementing the filter as described in Section 5.2.1. In this example the direct structure shown in Fig. 5.6 was chosen because it is especially easy to implement with the special multiply/accumulate instructions on the TMS32010. For this short filter there should be little problem with quantization effects if 16-bit coefficients are used and 32 bits are used to accumulate the output sum.

STEP 3. Next we must scale the unit-pulse response coefficients to limit overflow (see Section 5.3.2). Here a tradeoff must be made. If we scale the filter coefficients small enough so that overflow will never happen, the output signal to

```
1*********************************************************
            finite impulse response (fir)
           linear phase digital filter design
               remes exchange algorithm
                  bandpass filter

                filter length =  21

             ***** impulse response *****
         h( 1) =   0.18255439e-01 = h( 21)
         h( 2) =   0.55136755e-01 = h( 20)
         h( 3) =  -0.40910728e-01 = h( 19)
         h( 4) =   0.14930855e-01 = h( 18)
         h( 5) =   0.27568584e-01 = h( 17)
         h( 6) =  -0.59407797e-01 = h( 16)
         h( 7) =   0.44841841e-01 = h( 15)
         h( 8) =   0.31902660e-01 = h( 14)
         h( 9) =  -0.14972545e+00 = h( 13)
         h(10) =   0.25687239e+00 = h( 12)
         h(11) =   0.69994062e+00 = h( 11)
                    band  1           band  2
lower band edge     0.              0.3700000
upper band edge     0.3300000       0.5000000
desired value       1.0000000       0.
weighting           1.0000000       1.0000000
deviation           0.0988697       0.0988697
deviation in db     0.8189238     -20.0987320
```

FIGURE 5.13. Specifications for FIR design example (Example 3.14).

l_1 Scaled Coefficients

		Decimal		Hex
h_1	=	0.008697	=	011D
h_2	=	0.026267	=	035D
h_3	=	-0.019490	=	FD82
h_4	=	0.007113	=	00E9
h_5	=	0.013134	=	01AE
h_6	=	-0.028302	=	FC62
h_7	=	0.021363	=	02BC
h_8	=	0.015199	=	01F2
h_9	=	-0.071330	=	F6DF
h_{10}	=	0.122376	=	0FAA
h_{11}	=	0.333457	=	2AAF

FIGURE 5.14. Scaled coefficients for design example.

quantization noise ratio will be smaller than if another scaling strategy, which allows occasional overflow, is used. The l_1 and the l_2 norms were calculated for this example.

$$l_1 \text{ norm} = 2.09905, \qquad l_2 \text{ norm} = 0.831802.$$

Since we expected little trouble from quantization noise, we used the most conservative scaling strategy to guarantee that overflow would never occur. All coefficients were divided by the l_1 norm to give the scaled unit-pulse response listed in Fig. 5.14. To ensure that the frequency response is still acceptable with the quantized coefficients, one should plot it and compare it with the response from unquantized coefficients.

STEP 4. Finally, we write an assembly language program for the direct, nonrecursive implementation of the filter, following the program shown in Fig. 5.9. A complete assembly language program for the TMS32010 are in the appendix (Program 11).

REFERENCES

[1]   R. A. Roberts and C. T. Mullis, *Digital Signal Processing*, Reading, MA: Addison-Wesley, 1987.

[2]   L. B. Jackson, *Digital Filters and Signal Processing*, Boston: Kluwer, 1986.

[3]   L. R. Rabiner and B. Gold, *Theory and Application of Digital Signal Processing*, Englewood Cliffs, NJ: Prentice-Hall, 1975.

[4]   H. W. Schüssler, *Digital Systems for Signal Processing*, Berlin: Springer-Verlag, 1973. (in German).

[5]   N. S. Jayant and P. Noll, *Digital Coding of* Waveforms, Englewood Cliffs, NJ: Prentice-Hall, 1984.

[6]   *TMS32010 Users Guide*, Texas Instruments, 1985.

[7]   D. M. Kodek, "Design of Optimal Finite Wordlength FIR Digital Filters Using Integer Programming Techniques," *IEEE Trans. ASSP* **28**, 304–308 (1980).

[8]   D. M. Kodek and K. Steiglitz, "Comparison of Optimal and Local Search Methods for Designing Finite Worldlength FIR Digital Filters," *IEEE Trans. Circuits Systems* **28**, 28–32 (1981).

Part III

Infinite Impulse Response (IIR) Filters

6

Properties of Infinite Impulse-Response Filters

Digital filters with an infinite-duration impulse response (IIR) have character-istics that make them useful in many applications. This chapter develops and discusses the properties and characteristics of these filters.

Because of the feedback necessary in an implementation, the infinite impulse response (IIR) filter is also called a recursive filter or, sometimes, an autoregressive moving-average filter (ARMA). In contrast to the FIR filter with a polynomial transfer function, the IIR filter has a rational transfer function. The transfer function being a ratio of polynomials means it has finite poles as well as zeros, and the frequency-domain design problem becomes a rational function approximation problem in contrast to the polynomial approximation for the FIR filter. This gives considerably more flexibility and power, but brings with it certain problems in both design and implementation.[1-4]

The defining relationship between the input and output variables for the IIR filter is given by

$$y(n) = - \sum_{k=1}^{N} a(k)y(n-k) + \sum_{k=0}^{M} b(k)x(n-k) \tag{6.1}$$

The second summation in (6.1) is exactly the same moving average of the present plus past M values of the input that occurs in the definition of the FIR filter in (2.1). The difference arises from the first summation, which is a weighted sum of the previous N output values. This is the feedback or recursive part that causes the response to an impulse input theoretically to endure forever. The calculation of each output term $y(n)$ from (6.1) requires $N + M + 1$ multiplications and $N + M$ additions. Other algorithms or structures for calculating $y(n)$ may require more or less arithmetic. They are discussed in Chapter 8.

Just as in the case of the FIR filter, the output of an IIR filter can also be

153

calculated by convolution.

$$y(n) = \sum_{k=0}^{\infty} h(k)x(n - k). \tag{6.2}$$

In this case the duration of the impulse response $h(n)$ is infinite, and, therefore, the number of terms in (6.2) is infinite. The $N + M + 1$ operations required in (6.1) are clearly preferable to the infinite number required by (6.2). This gives a hint as to why the IIR filter is very efficient. The details will become clear as the characteristics of the IIR filter are developed in this chapter.

6.1 FREQUENCY-DOMAIN FORMULATION OF IIR FILTERS

The transfer function of a filter is defined as the ratio $Y(z)/X(z)$, where $Y(z)$ and $X(z)$ are the z transforms of the output $y(n)$ and input $x(n)$, respectively. It is also the z transform of the impulse response. Using the definition of the z transform in (2.4), we obtain the transfer function of the IIR filter defined in (6.1):

$$H(z) = \sum_{n=0}^{\infty} h(n)z^{-n}. \tag{6.3}$$

This transfer function is also the ratio of the z transforms of the $a(n)$ and $b(n)$ terms.

$$H(z) = \frac{\sum_{n=0}^{M} b(n)z^{-n}}{\sum_{n=0}^{N} a(n)z^{-n}} = \frac{B(z)}{A(z)}. \tag{6.4}$$

The frequency response of the filter, as shown in Section 1.2, is found by setting $z = e^{j\omega}$, which gives (6.3) the form

$$H(\omega) = \sum_{n=0}^{\infty} h(n)e^{-jn\omega}. \tag{6.5}$$

Recall that this form assumes a sampling rate of $T = 1$. To simplify notation, we use $H(\omega)$ rather than $H(e^{j\omega})$ to denote the frequency response.

This frequency-response function is complex valued and consists of a magnitude and a phase. Even though the impulse response is a function of the discrete variable n, the frequency response is a function of the continuous-frequency variable ω and is periodic with period 2π, as was shown for the FIR case in Section 2.1.

Unlike the FIR filter case, exactly linear phase is impossible for the IIR filter. In (2.18) and (2.23) we showed that linear phase is equivalent to symmetry of the impulse response. This equivalency is clearly impossible for the IIR filter with an impulse response that is zero for $n < 0$ and nonzero for n going to infinity.

The FIR linear-phase filter allowed us to remove the phase from the design process. The resulting problem was a real-valued approximation problem requiring the solution of linear equations. The IIR filter design problem is more complicated. Linear phase is not possible, and the equations to be solved are generally nonlinear. The most common technique is to approximate the magnitude of the transfer function and let the phase take care of itself. If the phase is important, it becomes part of the approximation problem, which then is often difficult to solve.

6.2 CALCULATION OF IIR FILTER FREQUENCY RESPONSE

As shown in Sections 2.2 and 2.2.2, L equally spaced samples of $H(\omega)$ can be approximately calculated by taking an L-length DFT of $h(n)$ given in (6.5). However, unlike for the FIR filter, this requires that the infinitely long impulse response be truncated to at least length L. A more satisfactory alternative is to use the DFT to evaluate the numerator and denominator of (6.4) separately rather than to approximately evaluate (6.3). We do this by appending $L - N$ zeros to the $a(n)$ and $L - M$ zeros to the $b(n)$ from (6.1) and by taking length-L DFTs of both to give

$$H\left(\frac{2\pi k}{L}\right) = \frac{\text{DFT}\{b(n)\}}{\text{DFT}\{a(n)\}}, \tag{6.6}$$

where the division is a termwise division of each of the L values of the DFTs as a function of k. This direct method of calculation is a straightforward and flexible technique that does not involve truncation of $h(n)$ and the resulting error. Even nonuniform spacing of the frequency samples can be achieved by altering the DFT defined in (2.7) as was suggested for the FIR filter. Because IIR filters are generally lower in order than FIR filters, direct use of the DFT is usually efficient enough, and use of the FFT is not necessary. Since the $a(n)$ and $b(n)$ do not generally have the symmetries of the FIR $h(n)$, the DFTs cannot be made real; therefore, the shifting and stretching techniques of Section 2.2.2 are not applicable.

An example of the frequency-response plot of a third-order elliptic function low-pass filter with transfer function

$$H(z) = \frac{z^3 + 1.07669z^2 + 1.07669z + 1}{z^3 - 0.84315z^2 + 0.90535z - 0.25211}$$

$$= \frac{(z + 1)(z^2 + 0.07669z + 1)}{(z - 0.34361)(z^2 - 0.49954z + 0.73370)} \tag{6.7}$$

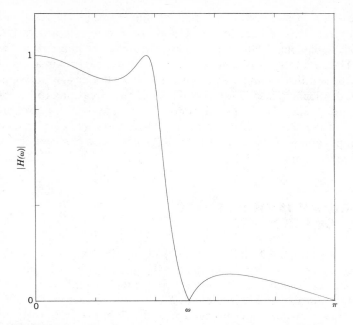

FIGURE 6.1. Magnitude frequency response of a third-order IIR filter.

is given in Fig. 6.1. The details for designing this filter are discussed in Section 7.2.8. A similar performance for the magnitude response would require a length of 18 for a linear-phase FIR filter.

6.3 LOCATIONS OF POLES AND ZEROS FOR IIR FILTERS

In Section 2.2.3 the possible locations of the zeros of the transfer function of an FIR linear-phase filter were analyzed. For the IIR filter there are poles as well as zeros. For most applications the coefficients $a(n)$ and $b(n)$ are real, and therefore the poles and zeros occur in complex-conjugate pairs, or they are real. A filter is stable if, for any bounded input, the output is bounded. This stability implies that the poles of the transfer function must be strictly inside the unit circle of the complex z plane. Indeed, the possibility of an unstable filter in IIR filter design is a serious problem that does not exist for FIR filters. An important characteristic of any design procedure is the guarantee of stable designs, and an important ability in the analysis of a given filter is the determination of stability. For a linear filter analysis, stability determination involves the zeros of the denominator polynomial of (6.4). The location of the zeros of the numerator, which are the zeros of $H(z)$, are important to the performance of the filter, but they have no effect on stability.

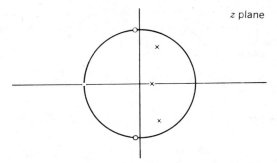

FIGURE 6.2. Pole and zero locations for a third-order IIR filter.

If all poles and zeros of a transfer function are inside or on the unit circle of the z plane, the filter is called *minimum phase*. The effects on the magnitude of the transfer function of a pole or a zero at a radius r from the origin of the z plane are exactly the same as a pole or zero at the same angle but at a radius of $1/r$. However, the effect on the phase characteristics is different. Because stable filters only are generally used in practice, all poles must be inside the unit circle. For a given magnitude response there are two possible locations for each zero not on the unit circle. The location that is inside gives the least phase shift—hence the name "minimum-phase" filter. The locations of the poles and zeros of the example in (6.7) are given in Fig. 6.2.

Since evaluating the frequency response of a transfer function is the same as evaluating $H(z)$ around the unit circle in the z plane, a comparison of the frequency-response plot in Fig. 6.1 and the pole-zero locations in Fig. 6.2 gives insight into the effects of pole and zero locations on the frequency response. When it is desirable to reject certain bands of frequencies, zeros of the transfer function will be located on the unit circle at locations corresponding to those frequencies. This case is illustrated in the examples in Chapter 7.

By using both poles and zeros to describe an IIR filter, we can do much more than in the FIR filter case where only zeros exist. Indeed, an FIR filter is a special case of an IIR filter with a zero-order denominator. This generality and flexibility do not come without a price. The poles are more difficult to realize than the zeros, and the design is more complicated.

Summary

This chapter gave the basic definition of the IIR or recursive digital filter and compared it to a generalization of the FIR filter described in previous chapters. The feedback terms in the IIR filter cause the transfer function to be a rational function with poles as well as zeros. This feedback and the resulting poles of the transfer function give a more versatile filter: fewer stored coefficients and less arithmetic are required. Unfortunately, it also destroys the possibility of linear phase and introduces the possibility of instability and greater sensitivity to the effects of quantization. The design methods, which are more complicated than for the FIR filter, are discussed in Chapter 7, and the implementation, which also is more complicated, is discussed in Chapter 8.

REFERENCES

[1]   A. V. Oppenheim and R. W. Schafer, *Digital Signal Processing*, Englewood Cliffs, NJ: Prentice-Hall, 1975.

[2]   L. R. Rabiner and B. Gold, *Theory and Application of Digital Signal Processing*, Englewood Cliffs, NJ: Prentice-Hall, 1975.

[3]   F. J. Taylor, *Digital Filter Design Handbook*, New York: Dekker, 1983.

[4]   R. A. Roberts and C. T. Mullis, *Digital Signal Processing*, Reading, MA: Addison-Wesley, 1987.

[5]   L. B. Jackson, *Digital Filters and Signal Processing*, Boston: Kluwer, 1986.

7

Design of Infinite Impulse-Response Filters

The design of a digital filter is usually specified in terms of the characteristics of the signals to be passed through the filter. In many cases the signals are described in terms of their frequency content. For example, even though it cannot be predicted just what a person may say, it can be predicted that the speech will have frequency content between 300 and 4000 Hz. Therefore, a filter can be designed to pass speech without knowing what the speech is. This frequency-domain description is true of many signals and of many types of noise or interference. For these reasons, among others, specifications for filters are generally given in terms of the frequency response of the filter.

The basic IIR filter design process is similar to that for the FIR problem:

1. Choose a desired response, usually in the frequency domain.
2. Choose an allowed class of filters—in this case, the Nth-order IIR filters.
3. Establish a measure of distance between the desired response and the actual response of a member of the allowed class.
4. Develop a method to find the best allowed filter as measured by being closest to the desired response.

This chapter develops several practical methods for IIR filter design. A very important set of methods is based on converting Butterworth, Chebyshev I and II, and elliptic-function analog filter designs to digital filter designs by both the impulse-invariant method and the bilinear transformation. The characteristics of these four approximations are based on combinations of a Taylor series and a Chebyshev approximation in the pass band and stop band. Many results from this chapter can be used for both analog and digital filter design.

Extensions of the frequency-sampling and LS error designs for the FIR filter are developed for the IIR filter. This chapter describes several direct iterative numerical methods for optimal approximation. Prony's method and direct numerical methods are presented for designing IIR filters according to time-domain specifications.

The discussion of the four classical low-pass filter design methods is arranged so that each method has a section on properties and a section on design procedures. There are also design programs in the appendix. An experienced person can simply use the design programs. A less-experienced designer should read the design procedure material, and someone who wants to understand the theory in order to modify the programs, develop new programs, or better understand the given ones should study the properties sections and consult the references.

7.1 RATIONAL FUNCTION APPROXIMATION

The mathematical problem inherent in the frequency-domain filter design problem is the approximation of a desired complex frequency-response function $H_d(z)$ by a rational transfer function $H(z)$ with an Mth-degree numerator and an Nth-degree denominator for values of the complex variable z along the unit circle of $z = e^{j\omega}$. This approximation is achieved by minimizing an error measure between $H_d(\omega)$ and $H(\omega)$.

For the digital filter design problem, the mathematics are complicated by the approximation being defined on the unit circle. In terms of z, frequency is a polar coordinate variable. It is often much easier and clearer to formulate the problem such that frequency is a rectangular coordinate variable, which is the way it naturally occurs for analog filters using the Laplace complex variable s. A particular change of complex variable that converts the polar coordinate variable to a rectangular coordinate variable is the bilinear transformation[1,2]

$$z = \frac{s + 1}{s - 1}. \tag{7.1}$$

The details of the bilinear and alternative transformations are covered in Section 7.3. For the purposes of this section it is sufficient for us to observe[18,19] that the frequency response of a filter in terms of the new variable is found by evaluating $H(s)$ along the imaginary axis (i.e., for $s = j\omega$). The frequency response of analog filters is obtained in exactly this way.

There are two reasons that the approximation process is often formulated in terms of the square of the magnitude of the transfer function rather than in terms of the real and/or imaginary parts of the complex transfer function or in terms of the magnitude of the transfer function. The first reason is that the squared magnitude frequency-response function is an analytic, real-valued function of a real variable, and this considerably simplifies the problem of finding a "best" solution. The second reason is that the effects of the signal or interference are often stated in terms of the energy or power, which is proportional to the square of the magnitude of the signal or noise.

To move back and forth between the transfer function $F(s)$ and the squared magnitude frequency response $|F(j\omega)|^2$, we define an intermediate function. We define the analytic complex-valued function of the complex variable s by

$$\mathscr{F}(s) = F(s)F(-s), \tag{7.2}$$

which is related to the squared magnitude by

$$\mathscr{F}(s)\bigg|_{s=j\omega} = |F(j\omega)|^2. \tag{7.3}$$

If

$$F(j\omega) = R(\omega) + jI(\omega),$$

then

$$\begin{aligned} |F(j\omega)|^2 &= R(\omega)^2 + I(\omega)^2 \\ &= (R(\omega) + jI(\omega))(R(\omega) - jI(\omega)) \\ &= F(s)F(-s)\bigg|_{s=j\omega}. \end{aligned}$$

In this context the approximation is arrived at in terms of $F(j\omega)$, and the result is an analytic function $\mathscr{F}(s)$ with a factor $F(s)$, which is the desired filter transfer function in terms of the rectangular variable s. We can define a comparable function in terms of the digital transfer function, using the polar variable z by defining

$$\mathscr{H}(z) = H(z)H\left(\frac{1}{z}\right), \tag{7.4}$$

which gives the magnitude squared frequency response when evaluated around the unit circle—that is, $z = e^{j\omega}$.

The next section develops four useful approximations, using the continuous-time Laplace transform formulation in s. These approximations will be transformed into digital transfer functions by techniques covered in Section 7.3. They can also be used directly for analog filter design.

7.2 CLASSICAL ANALOG LOW-PASS FILTER APPROXIMATIONS

Four basic filter approximations are considered to be standard. They are often developed and presented in terms of a normalized low-pass filter that can be modified to give other versions, such as high-pass or bandpass filters. These four forms use Taylor series approximations and Chebyshev approximations in various combinations.[1,3,7,17-19] None is defined in terms of a mean squared error measure. Although it would be an interesting error criterion, the reason is that there is no closed-form solution to the LS error approximation problem, which is nonlinear for the IIR filter.

This section develops the four classical approximations in terms of the Laplace transform variable s. They can be used as prototype filters to be converted into digital filters or used directly for analog filter design.

The desired low-pass filter frequency response is similar to the case for the FIR filter, given in Fig. 3.1 and (3.26). Here it is expressed in terms of the magnitude squared of the transfer function, which is a function of $s = j\omega$ and is illustrated in Fig. 7.1.

The Butterworth filter uses a Taylor series approximation to the ideal at both $\omega = 0$ and $\omega = \infty$. The Chebyshev filter uses a Chebyshev (min-max) approximation across the pass band and a Taylor series at $\omega = \infty$. The inverse or type II Chebyshev filter uses a Taylor series approximation at $\omega = 0$ and a Chebyshev approximation across the stop band. The elliptic function filter uses a Chebyshev approximation across both the pass band and stop band. The squared magnitude frequency response for these approximations to the ideal in Fig. 7.1 is given in Fig. 7.2, and the design is developed in the following sections.

7.2.1 Butterworth Filter Properties

This section develops the properties of the Butterworth filter, which has as its basic concept a Taylor series approximation to the desired frequency response. The measure of the approximation is the number of terms in the Taylor series expansion of the actual frequency response that can be made equal to those of the desired frequency response. The optimal or best solution will have the maximum number of terms equal. The Taylor series is a power series expansion of a function in the form

$$F(\omega) = K_0 + K_1\omega + K_2\omega^2 + K_3\omega^3 + \cdots,$$

where

$$K_0 = F(0), \quad K_1 = \frac{dF(\omega)}{d\omega}\bigg|_{\omega=0}, \quad K_2 = \frac{1}{2}\frac{d^2F(\omega)}{d\omega^2}\bigg|_{\omega=0},$$

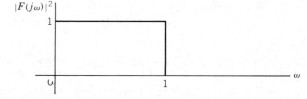

FIGURE 7.1. Desired frequency response of an ideal low-pass filter.

and so on, with the coefficients of the Taylor series being proportional to the various order derivatives of $F(\omega)$ evaluated at $\omega = 0$. A basic characteristic of this approach is that the approximation is all performed at one point (i.e., at one frequency). The ability of this approach to give good results over a range of frequencies depends on the analytic properties of the response.

The general form for the squared magnitude response is an even function of ω and, therefore, is a function of ω^2 expressed as

$$\mathscr{F}(j\omega) = \frac{d_0 + d_2\omega^2 + d_4\omega^4 + \cdots + d_{2M}\omega^{2M}}{c_0 + c_2\omega^2 + c_4\omega^4 + \cdots + c_{2N}\omega^{2N}}. \tag{7.5}$$

To obtain a solution that is a low-pass filter, we perform the Taylor series expansion around $\omega = 0$, requiring $\mathscr{F}(0) = 1$ and $\mathscr{F}(j\infty) = 0$ (i.e., $d_0 = c_0$, $N > M$, and $c_{2N} \neq 0$). We write it as

$$\mathscr{F}(j\omega) = 1 + E(\omega). \tag{7.6}$$

Combining (7.5) and (7.6) gives

$$d_0 + d_2\omega^2 + \cdots + d_{2M}\omega^{2M} = c_0 + c_2\omega^2 + \cdots + c_{2N}\omega^{2N}$$
$$+ E(\omega)[c_0 + c_2\omega^2 + \cdots]. \tag{7.7}$$

The best Taylor approximation requires that $\mathscr{F}(j\omega)$ and the desired ideal response have as many terms as possible equal in their Taylor series expansion at a given frequency. For a low-pass filter the expansion is around $\omega = 0$, which requires $E(\omega)$ to have as few low-order ω terms as possible. This condition is achieved by setting

$$\begin{aligned}
c_0 \quad &= d_0, \\
c_2 \quad &= d_2, \\
&\vdots \\
c_{2M} \quad &= d_{2M}, \\
c_{2M+2} &= 0, \\
&\vdots \\
c_{2N-2} &= 0, \\
c_{2N} \quad &= \text{nonzero.}
\end{aligned} \tag{7.8}$$

Because the ideal response in the pass band is a constant, the Taylor series approximation is often called *maximally flat*.

Equation (7.8) states that the numerator of the transfer function may be chosen arbitrarily. Then by setting the denominator coefficients of $\mathscr{F}(s)$ equal to the numerator coefficients plus one higher-order term, we obtain an optimal Taylor's series approximation[19].

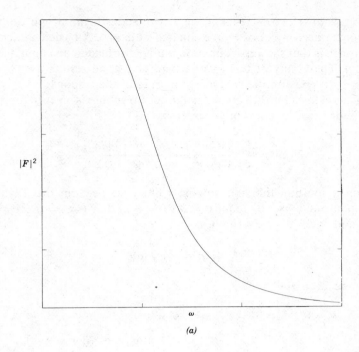

$|F|^2$

ω

(a)

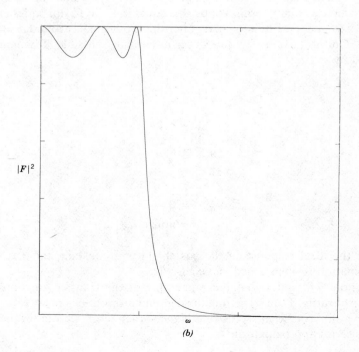

$|F|^2$

ω

(b)

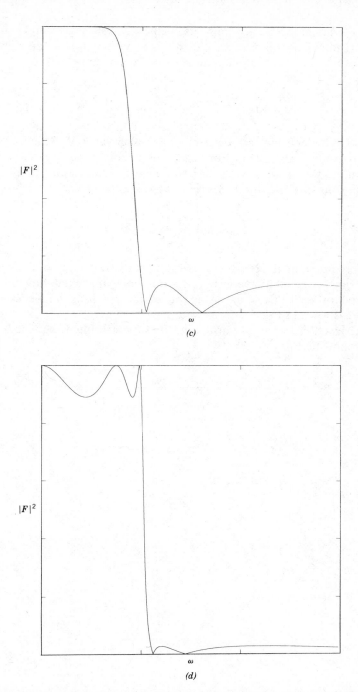

$|F|^2$

ω

(c)

$|F|^2$

ω

(d)

FIGURE 7.2. Frequency responses of the four classical low-pass IIR filter approximations. (*a*) Butterworth; (*b*) Chebyshev; (*c*) inverse Chebyshev; (*d*) elliptic function.

165

Since the numerator is arbitrary, its coefficients can be chosen for a Taylor approximation to zero at $\omega = \infty$. We do this by setting $d_0 = 1$ and all other d's equal to zero. The resulting magnitude squared function is[1,19]

$$\mathscr{F}(j\omega) = \frac{1}{1 + c_{2N}\omega^{2N}}.$$

The value of the constant c_{2N} determines at which value of ω the transition of pass band to stop band occurs. For this development it is normalized to $c_{2N} = 1$, which causes the transition to occur at $\omega = 1$. These approximations and normalizations give the simple form for what is called the *Butterworth filter*:

$$\mathscr{F}(j\omega) = \frac{1}{1 + \omega^{2N}}. \tag{7.9}$$

This approximation is sometimes called maximally flat at both $\omega = 0$ and $\omega = \infty$, since it is simultaneously a Taylor series approximation to unity at $\omega = 0$ and to zero at $\omega = \infty$. A graph of the resulting frequency-response function is shown in Fig. 7.3 for several N.

The characteristics of the normalized Butterworth filter frequency response are the following:

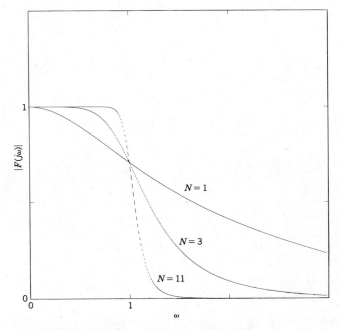

FIGURE 7.3. Butterworth filter frequency responses.

1. It is very close to the ideal near $\omega = 0$ and $\omega = \infty$.
2. It is very smooth at all frequencies with a monotonic decrease from $\omega = 0$ to ∞.
3. The largest difference occurs between the ideal and actual responses near the transition at $\omega = 1$, where $|F(j_1)|^2 = \frac{1}{2}$.

Although not part of the approximation addressed, the phase curve is also very smooth.

An important feature of the Butterworth filter is the closed-form formula for the solution, $F(s)$. From (7.3) the expression for $\mathcal{F}(s)$ may be determined as

$$F(s)F(-s) = \frac{1}{1 + (-s^2)^N}. \tag{7.10}$$

This function has $2N$ poles evenly spaced around a unit radius circle and $2N$ zeros at infinity. The determination of $F(s)$ is very simple. To have a stable filter, we select $F(s)$ to have the N left-hand plane poles and N zeros at infinity; $F(-s)$ will necessarily have the right-hand plane poles and the other N zeros at infinity. The locations of these poles on the complex s plane for $N = 1, 2, 3,$ and 4 are shown in Fig. 7.4.

Pole Location
Because of the geometry of the pole positions, simple formulas are easy to derive for the pole locations. If the real and imaginary parts of the pole location are denoted by

$$s = u + j\omega,$$

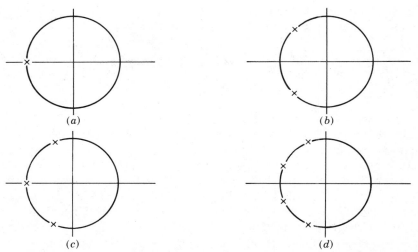

FIGURE 7.4. Pole locations for Butterworth filter transfer. Functions $F(s)$ on the complex s plane. (a) $N = 1$; (b) $N = 2$; (c) $N = 3$; (d) $N = 4$.

the locations of the N poles are given by

$$u_k = -\cos\left(\frac{k\pi}{2N}\right), \qquad \omega_k = \sin\left(\frac{k\pi}{2N}\right) \tag{7.11}$$

for N values of k where

$$k = \pm 1, \ \pm 3, \ \pm 5, \ldots, \ \pm(N-1) \quad \text{for } N \text{ even,}$$
$$k = \quad 0, \ \pm 2, \ \pm 4, \ldots, \ \pm(N-1) \quad \text{for } N \text{ odd.}$$

Because the coefficients of the numerator and denominator polynomials of $F(s)$ are real, the roots occur in complex-conjugate pairs. The conjugate pairs in (7.11) can be combined to be the roots of second-order polynomials so that for N even $F(s)$ has the partially factored form

$$F(s) = \prod_k \frac{1}{s^2 + 2\cos(k\pi/2N)s + 1} \tag{7.12a}$$

for $k = 1, 3, 5, \ldots, N - 1$. For N odd, $F(s)$ has a single real pole; therefore

$$F(s) = \frac{1}{s + 1} \prod_k \frac{1}{s^2 + 2\cos(k\pi/2N)s + 1} \tag{7.12b}$$

for $k = 2, 4, 6, \ldots, N - 1$. This form is convenient for the cascade and parallel realizations discussed in Chapter 8.

A single formula for the pole locations for both even and odd N is

$$u_k = -\sin\left(\frac{(2k + 1)\pi}{2N}\right), \qquad \omega_k = \cos\left(\frac{(2k + 1)\pi}{2N}\right) \tag{7.13}$$

for N values of k, where $k = 0, 1, 2, \ldots, N - 1$.

One of the important features of the Butterworth filter design formulas is that the pole locations are found by independent calculations, which do not depend on each other or on factoring a polynomial. Program 9 calculates these values.

The classical form of the Butterworth filter given in (7.10) is discussed in many books.[1,3,7,11,12,18,19] The less well-known form given in (7.8) also has many useful applications[19]. If the frequency location of unwanted signals is known, the zeros of the transfer function given by the numerator can be set to best reject them. It is then possible to choose the pole by using (7.8) to have a pass band as flat as the classical Butterworth filter. Unfortunately, there are no formulas for the pole locations; therefore, the denominator polynomial must be factored.

Summary

This section derived design procedures and formulas for a class of filter transfer functions that approximate the ideal desired frequency response by a Taylor series. If the approximation is made at $\omega = 0$ and $\omega = \infty$, the resulting filter is called a Butterworth filter and the response is called maximally flat at zero and infinity. This filter has a very smooth frequency response and, although not explicitly designed for, a smooth phase response. Simple formulas for the pole locations were derived and are implemented in the design program in the appendix.

7.2.2 Butterworth Filter Design Procedures

This section considers the process of going from given specifications to use of the approximation results derived in the previous section. The Butterworth filter is the simplest of the four classical filters in that all the approximation effort is placed at two frequencies: $\omega = 0$ and $\omega = \infty$. The transition from pass band to stop band occurs at a normalized frequency $\omega = 1$. Assuming that this transition frequency or band edge can later be scaled to any desired frequency, the only parameter to be chosen in the design process is the order N.

The filter specifications that are consistent with what is optimized in the Butterworth filter are the degree of "flatness" at $\omega = 0$ (DC) and at $\omega = \infty$. The higher the order, the flatter the frequency response at these two points. Because of the analytic nature of rational functions, the flatter the response is at $\omega = 0$ and ∞, the closer it stays to the desired response throughout the whole pass band and stop band. An indirect consequence of the filter order is the slope of the response at the transition between pass band and stop band. The slope of the squared magnitude frequency response at $\omega = 1$ is

$$\text{slope} = \mathscr{F}'(j1) = -\frac{N}{2}. \tag{7.14}$$

The effects of the increased flatness and increased transition slope of the frequency response as N increases are illustrated in Fig. 7.3.

In some cases specifications state the response must stay above or below a certain value over a given frequency band. Although this type of specification is more compatible with a Chebyshev error optimization, it is possible to design a Butterworth filter to meet the requirements. If the magnitude of the frequency response of the filter over the pass band of $0 < \omega < \omega_p$ must remain between unity and G, where $\omega_p < 1$ and $G < 1$, we find the required order by determining the smallest integer N satisfying

$$N \geqslant \frac{\log((1/G)^2 - 1)}{2 \log \omega_p}. \tag{7.15}$$

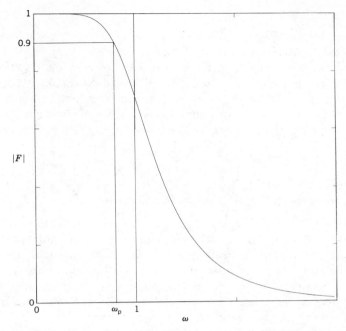

FIGURE 7.5. Pass-band specifications for designing a Butterworth filter.

This specification is illustrated in Fig. 7.5, where $|F|$ must remain above 0.9 for ω up to 0.9; that is, $G = 0.9$ and $\omega_p = 0.9$. These requirements require an order of at least $N = 7$.

   If stop-band performance is stated in the form of requiring that the response stay below a certain value for frequency above a certain value—that is, $|F| < G$ for $\omega > \omega_s$—the order is determined by the same formula (7.15) with ω_p replaced by ω_s.

Example 7.1. Design of a Butterworth Low-Pass IIR Filter
   To illustrate the calculations, we design a low-pass Butterworth filter. We want the frequency response to stay above 0.8 for frequencies up to 0.9. Formula (7.15) for determining the order gives a value of 2.73; therefore, the order is 3. The analytic function corresponding to the squared magnitude frequency response in (7.10) is

$$F^2(j\omega) = \frac{1}{1 + \omega^6}.$$

The transfer function corresponding to the left half-plane poles of $F'(s)$ are calculated from (7.11) or (7.12) to give

$$F(s) = \frac{1}{(s + 1)(s + 0.5 + j0.866)(s + 0.5 - j0.866)}, \tag{7.16}$$

$$F(s) = \frac{1}{(s + 1)(s^2 + s + 1)},\tag{7.17}$$

$$F(s) = \frac{1}{s^3 + 2s^2 + 2s + 1}.\tag{7.18}$$

We obtain the frequency response by setting $s = j\omega$, which has a plot illustrated in Fig. 7.3 for $N = 3$. The pole locations are the same as shown in Fig. 7.4c.

7.2.3 Chebyshev Filter Properties

Frequently the Butterworth filter does not give a sufficiently good approximation across the complete pass band. The Taylor series approximation is often not suited to the way specifications are given for filters. An alternative error measure is the maximum of the absolute value of the difference between the actual filter response and the ideal response. This measure is considered over the total pass band. It is the Chebyshev error measure, which was defined and applied to the FIR filter design problem in Section 3.3. For the IIR filter the Chebyshev error is minimized over the pass band, and a Taylor series approximation at $\omega = \infty$ is used to determine the stop-band performance. This mixture of methods in the IIR case is called the *Chebyshev filter*, and we obtain simple design formulas just as for the Butterworth filter.

The design of Chebyshev filters is particularly interesting, because the results of a very elegant theory ensure that constructing a frequency-response function with the proper form of equal ripple in the error will give a minimum Chebyshev error without explicitly minimizing anything. That allows a straightforward set of design formulas to be derived, which can be viewed as a generalization of the Butterworth formulas.[18,19]

The form for the magnitude squared of the frequency-response function for the Chebyshev filter is

$$|F(j\omega)|^2 = \frac{1}{1 + \varepsilon^2 C_N^2(\omega)},\tag{7.19}$$

where $C_N(\omega)$ is an Nth-order Chebyshev polynomial and ε is a parameter that controls the ripple size. This polynomial in ω has very special characteristics that result in the optimality of the response function (7.19).

Chebyshev Polynomials
The Chebyshev polynomial is a powerful function in approximation theory. Although the function is a polynomial, it is best defined and developed in terms of trigonometric functions by[1,7,18,19]

$$C_N(\omega) = \cos(N \cos^{-1}(\omega)),\tag{7.20}$$

where $C_N(\omega)$ is an Nth-order, real-valued function of the real variable ω. The development is made clearer by introducing an intermediate complex variable ϕ:

$$C_N(\omega) = \cos(N\phi), \qquad \text{where } \omega = \cos(\phi). \qquad (7.21)$$

Although this definition of $C(\omega)$ may not at first appear to give a polynomial, the following recursive relation derived from (7.21) shows that it is indeed a polynomial.

$$C_{N+1}(\omega) = 2\omega C_N(\omega) - C_{N-1}(\omega). \qquad (7.22)$$

From (7.20) it is clear that $C_0 = 1$ and $C_1 = \omega$, and from (7.22) it follows that

$$C_2 = 2\omega^2 - 1,$$
$$C_3 = 4\omega^3 - 3\omega,$$
$$C_4 = 8\omega^4 - 8\omega^2 + 1,$$
$$\vdots \qquad (7.23)$$

Other relations useful for developing these polynomials are

$$C_N^2 = \tfrac{1}{2}[C_{2N} + 1],$$
$$C_{MN} = C_M(C_N(\omega)) \qquad \text{where } M \text{ and } N \text{ are coprime.} \qquad (7.24)$$

These functions are remarkable[18,19]. They oscillate between $+1$ and -1 for $-1 < \omega < 1$ and go monotonically to $\pm\infty$ outside that domain. All N of their zeros are real and fall in the domain $-1 < \omega < 1$; that is, C_N approximates zero over the range of ω from -1 to $+1$. In addition, the values for ω where C_N reaches its local maxima and minima and is zero are easily calculated from

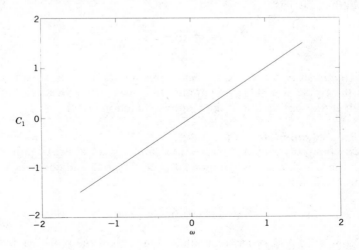

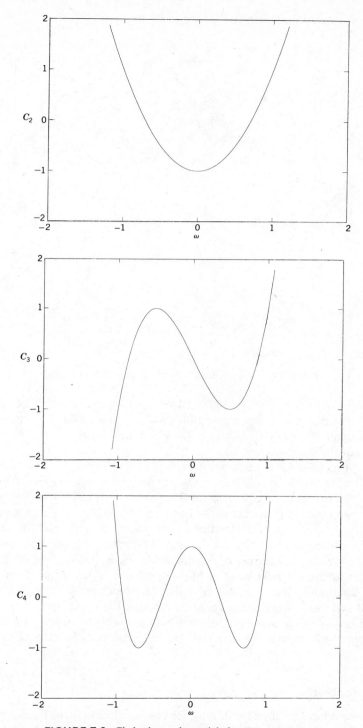

FIGURE 7.6. Chebyshev polynomials for $N = 1, 2, 3$, and 4.

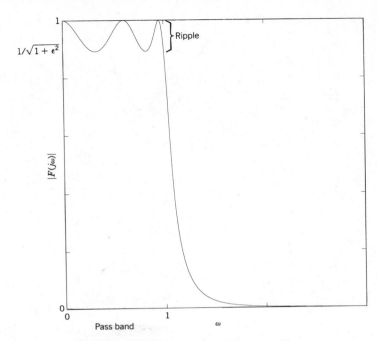

FIGURE 7.7. Fifth-order Chebyshev filter frequency response.

(7.21). For $-1 < \omega < 1$ we can plot $C_N(\omega)$ by using the concept of Lissajous figures[18]. Figure 7.6 shows example plots for C_1, C_2, C_3, and C_4. Figure 7.7 gives the filter frequency-response function for $N = 5$ and shows the pass-band ripple in terms of the parameter ε.

The approximation parameters must be clearly understood. The pass-band ripple is defined to be the difference between the maximum and the minimum of $|F|$ over the pass-band frequencies of $0 < \omega < 1$. This point can be confusing because two definitions appear in the literature. Most digital[1,2] and analog[18,19] filter design books use the definition just stated. Approximation literature, especially concerning FIR filters, and the ASPI design program[11] use one half of this value, which is a measure of the maximum error, $\big|\,|F| - |F_d|\,\big|$, where $|F_d|$ is the center line in the pass band of Fig. 7.7, around which $|F|$ oscillates.

The Chebyshev theory states that the maximum error over that band is minimal and this optimal approximation function has equal ripple over the pass band. It is easy to see that ε in (7.19) determines the ripple in the pass band and the order N determines the rate that the response goes to zero as ω goes to infinity.

Pole Locations

We now develop a method for finding the pole locations for the Chebyshev filter transfer function. The details of this section can be skipped, and the results in (7.31) can be used if the reader desires.

From (7.19) we see that the poles of $\mathscr{F}(s)$ occur when

$$1 + \varepsilon^2 C_N^2\left(\frac{s}{j}\right) = 0 \tag{7.25}$$

or

$$C_N = \pm\frac{j}{\varepsilon}.$$

From (7.21) define $\phi = \cos^{-1}(\omega)$ with real and imaginary parts given by

$$\phi = \cos^{-1}(\omega) = u + jv. \tag{7.26}$$

This gives, from (7.21) and (7.25),

$$C_N = \cos(N\phi) = \cos(Nu)\cosh(Nv) - j\,\sin(Nu)\sinh(Nv) = \pm\frac{j}{\varepsilon}, \tag{7.27}$$

which implies that the real part of C_N is zero. Therefore

$$\cos(Nu)\cosh(Nv) = 0,$$

which implies

$$\cos(Nu) = 0,$$

which implies that u assumes values of

$$u = u_k = \frac{(2k+1)\pi}{2N}, \qquad k = 0, 1, \ldots, N-1. \tag{7.28}$$

For these values of u $\sin(Nu) = \pm 1$, and (7.27) becomes

$$\sinh(Nv) = \frac{1}{\varepsilon},$$

which requires v to assume a value of

$$v = v_0 = \frac{\sinh^{-1}(1/\varepsilon)}{N}. \tag{7.29}$$

Using $s = j\omega$ and (7.26) give

$$s = j\omega = j\,\cos(\phi) = j\,\cos(u + jv)$$

$$= j\,\cos\left(\frac{(2k+1)\pi}{2N} + jv_0\right). \tag{7.30}$$

This equation gives the location of the N poles in the s plane as

$$s_k = \sigma_k + j\omega_k,$$

where

$$\sigma_k = -\sinh(v_0)\cos\left(\frac{k\pi}{2N}\right),$$

$$\omega_k = \cosh(v_0)\sin\left(\frac{k\pi}{2N}\right), \tag{7.31}$$

for N values of k, where

$$k = \begin{cases} \pm1, \pm3, \pm5, \ldots, \pm(N-1) & \text{for } N \text{ even,} \\ 0, \pm2, \pm4, \ldots, \pm(N-1) & \text{for } N \text{ odd.} \end{cases}$$

We can derive a partially factored form for $F(s)$ from (7.31) by using the same approach as for (7.12) for the Butterworth filter. For N even the form is

$$F(s) = \prod_k \frac{1}{s^2 - 2\sigma_k s + (\sigma_k^2 + \omega_k^2)} \tag{7.32a}$$

For $k = 1, 3, 5, \ldots, N-1$. For N odd, $F(s)$ has a single real pole, and therefore the form

$$F(s) = \frac{1}{s + \sinh(v_0)} \prod_k \frac{1}{s^2 - 2\sigma_k s + (\sigma_k^2 + \omega_k^2)} \tag{7.32b}$$

for $k = 2, 4, 6, \ldots, N-1$. This form is convenient for the cascade and parallel realizations discussed in Chapter 8.

A single formula for even and odd N is

$$\sigma_k = -\sinh(v_0)\sin\left(\frac{(2k+1)\pi}{2N}\right),$$

$$\omega_k = \cosh(v_0)\cos\left(\frac{(2k+1)\pi}{2N}\right) \tag{7.33}$$

for N values of k, where $k = 0, 1, 2, \ldots, N-1$. Note the similarity to the pole locations for the Butterworth filter in (7.12) and (7.13). Cross-multiplying, squaring, and adding the terms in (7.33) gives

$$\left(\frac{\sigma_k}{\sinh(v_0)}\right)^2 + \left(\frac{\omega_k}{\cosh(v_0)}\right)^2 = 1. \tag{7.34}$$

This equation is that of an ellipse and shows that the poles of a Chebyshev filter lie on an ellipse in a way similar to the way poles of a Butterworth filter lie on a circle.[1-3,7,18,19]

Summary

This section developed the classical Chebyshev filter approximation, which minimizes the maximum error over the pass band and uses a Taylor series approximation at infinity. Thus the error is equal ripple in the pass band. The transfer function was developed in terms of the Chebyshev polynomial, and explicit formulas were derived for the location of the transfer function poles. These formulae can be expressed as a modification of the pole locations for the Butterworth filter and are implemented in Program 9.

It is possible to develop a theory for Chebyshev pass-band approximation and arbitrary zero location similar to the Taylor series result in (7.8). That theory is described in references 24 and 25 and is not covered in this book.

7.2.4 Chebyshev Filter Design Procedures

The Chebyshev filter has a pass band optimized to minimize the maximum error over the complete pass-band frequency range, and a stop band controlled by the frequency response being maximally flat at $\omega = \infty$. The pass-band ripple and the filter order are the two parameters to be determined by the specifications.

The form for the specifications that is most consistent with the Chebyshev filter formulation is a maximum allowed error in the pass band and a desired degree of "flatness" at $\omega = \infty$. The slope of the response near the transition from pass band to stop band at $\omega = 1$ becomes steeper as both the order increases and the allowed pass-band and error ripple increases. The dropoff is more rapid than for the Butterworth filter.[18,19]

As stated earlier, the design parameters must be clearly understood to obtain a desired result. The pass-band ripple is defined to be the difference between the maximum and the minimum of $|F|$ over the pass-band frequencies of $0 \leqslant \omega \leqslant 1$. This point can be confusing because two definitions appear in the literature. Most digital[1,2] and analog[18,19] filter design books use the definition just stated. Approximation literature, especially concerning FIR filters, and the ASPI design program[11] use half this value, which is a measure of the maximum error, $||F| - |F_d||$, where $|F_d|$ is the center line in the pass band of Fig. 7.7 around which $|F|$ oscillates. The following formulas relate the pass-band ripple d, the pass-band ripple a in positive dB, and the transfer function parameter ε.

$$a = 10 \log(1 + \varepsilon^2) = -20 \log(1 - d), \tag{7.35}$$

$$\varepsilon = \sqrt{\frac{2d - d^2}{1 - 2d + d^2}} = \sqrt{10^{a/10} - 1}, \tag{7.36}$$

$$d = 1 - 10^{-a/20} = 1 - \frac{1}{\sqrt{1 + \varepsilon^2}}. \tag{7.37}$$

In some cases stop-band performance is not given in terms of degree of flatness at $\omega = \infty$, but in terms of a maximum allowed magnitude G in the stop band above a certain frequency ω_s; that is, $G > |F| > 0$ for $1 < \omega_s < \omega < \infty$. For a given ε this will determine the order as the smallest positive integer satisfying

$$N \geqslant \frac{\cosh^{-1}\left(\dfrac{\sqrt{1 - G^2}}{\varepsilon G^2}\right)}{\cosh^{-1}(\omega_s)} \tag{7.38}$$

The design of a Chebyshev filter involves the following steps:

1. The maximum allowed pass-band variation must be in the form of d or a. From this variation the parameter ε is calculated by (7.36).
2. The order N is determined by the desired flatness at $\omega = \infty$ or a maximum allowed response for frequencies above ω_s by (7.38).
3. v_0 is calculated from ε and n by (7.29), and the scale factors $\sinh(v_0)$ and $\cosh(v_0)$ are then determined.
4. The pole locations are calculated from (7.31) or (7.33) by scaling the poles of a Butterworth prototype filter.
5. These pole locations are combined in (7.32) to give the final filter transfer function.

This process is easily programmed for computer-aided design, as illustrated in Program 9 in the appendix.

If the design procedure uses (7.38) to determine the order and the right-hand side of the equation is not exactly an integer, it is possible to improve on the specifications. Direct use of the order with ε from (7.36) gives a stop-band gain at ω_s that is less than G, or the same design can be viewed as giving the maximum allowed gain G at a lower frequency than ω_s. An alternative approach is to solve (7.38) for a new value of ε, then cause (7.38) to be an equation with the specified ω_s and G. This approach gives a filter that exactly meets the stop-band specifications and gives a smaller pass-band ripple than originally requested. A similar set of alternatives exists for the elliptic function filter in Section 7.2.7.

Example 7.2. Design of a Chebyshev Low-Pass Filter
    The design specifications require a maximum pass-band ripple of $d = 0.1$ or $a = 0.91515$ dB and can allow no greater response than $G = 0.2$ for frequencies above $\omega_s = 1.6$ rad/s. Given $d = 0.1$ or $a = 0.91515$, equation (7.36) implies

$$\varepsilon = 0.484322. \tag{7.39}$$

Given $G = 0.2$ and $\omega_s = 1.6$, equation (7.38) implies an order of $N = 3$. From ε and N, v_0 is 0.49074 from (7.29) and

$$\sinh(v_0) = 0.510675, \qquad \cosh(v_0) = 1.122849. \tag{7.40}$$

These multipliers are used to scale the root locations of the third-order Butterworth filter in Example 7.1 to give

$$F(s) = \frac{1}{(s + 0.51067)(s + 0.25534 + j0.97242)(s + 0.25534 - j0.97242)},$$
(7.41)

$$F(s) = \frac{1}{(s + 0.510675)(s^2 + 0.510675s + 1.010789)},$$

$$F(s) = \frac{1}{s^3 + 1.02135s^2 + 1.271579s + 0.516185}.$$
(7.42)

The frequency response is shown in Fig. 7.8, and the pole locations on the s plane are shown in Fig. 7.9.

7.2.5 Inverse Chebyshev Filter Properties

A second form of the mixture of the Chebyshev approximation and a Taylor series approximation is called the *inverse Chebyshev filter* or the *Chebyshev II filter*. This error measure uses a Taylor's series for the pass band, just as for the

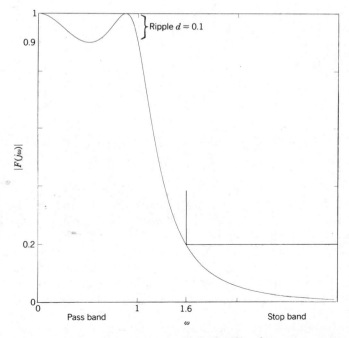

FIGURE 7.8. Example third-order Chebyshev filter frequency response.

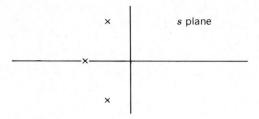

FIGURE 7.9. Pole locations in the s plane for the example Chebyshev filter.

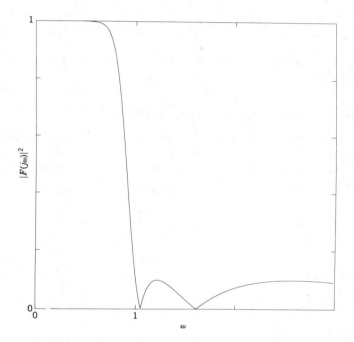

FIGURE 7.10. An inverse Chebyshev filter frequency response.

Butterworth filter, and minimizes the maximum error over the total stop band. It reverses the types of approximation used in the preceding section. A fifth-order example is illustrated in Fig. 7.10.

It is easier to modify the results from the regular Chebyshev filter than to develop the approximation directly. First, the frequency variable ω in the regular Chebyshev filter, described in (7.19), is replaced by $1/\omega$, which inter-changes the characteristics at $\omega = 0$ and $\omega = 0$ and does not change the performance at $\omega = 1$. Thus a Chebyshev low-pass filter is converted to a Chebyshev high-pass filter, as illustrated in Fig. 7.11. This high-pass character-

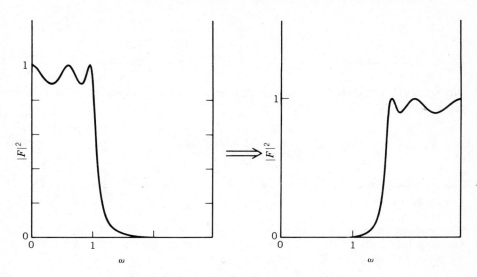

FIGURE 7.11. Low-Pass to high-pass transformation.

istic is subtracted from unity to give the desired low-pass inverse Chebyshev frequency response illustrated in Fig. 7.10. The resulting magnitude squared frequency-response function is

$$\mathscr{F}(j\omega) = \frac{\varepsilon^2 C_N^2(1/\omega)}{1 + \varepsilon^2 C_N^2(1/\omega)}. \tag{7.43}$$

Zero Locations
The zeros of the Chebyshev polynomial $C_N(\omega)$ are easily found from (7.20) by

$$C_N(\omega) = 0 \Rightarrow N \cos^{-1}(\omega) = \frac{(2k+1)\pi}{2}, \tag{7.44}$$

which requires

$$\omega_k = \cos\left(\frac{(2k+1)\pi}{2N}\right) \qquad \text{for } k = 0, 1, \ldots, N-1,$$

or

$$\omega_k = \sin\left(\frac{k\pi}{2N}\right) \tag{7.45}$$

for

$$k = \begin{cases} 0, \pm 2, \pm 4, \ldots, \pm(N-1), & N \text{ odd}, \\ \pm 1, \pm 3, \pm 5, \ldots, \pm(N-1), & N \text{ even}. \end{cases} \quad (7.46)$$

The zeros of the inverse Chebyshev filter transfer function are derived from (7.43) and (7.45) to give

$$\omega_{zk} = \frac{1}{\cos(2k+1)\pi/2N} \text{ or } \frac{1}{\sin(k\pi/2N)}. \quad (7.47)$$

The zero locations are not functions of ε; that is, they are independent of the stop-band ripple.

Pole Locations

The pole locations are the reciprocal of those for the regular Chebyshev filter given in (7.31) or (7.33). If the poles for the inverse filter are denoted by

$$s'_k = \sigma'_k + j\omega'_k,$$

the locations in terms of the variables of (7.32) or (7.33) are

$$\sigma'_k = \frac{\sigma_k}{\sigma_k^2 + \omega_k^2}, \qquad \omega'_k = \frac{\omega_k}{\sigma_k^2 + \omega_k^2}. \quad (7.48)$$

Although this gives a straightforward formula for calculating the location of the poles and zeros of the inverse Chebyshev filter, they do not lie on a simple geometric curve as did those for the Butterworth and Chebyshev filters. Note that the conditions of (7.8) for a Taylor series approximation with preset zero locations are satisfied in (7.43).

A partially factored form analogous to (7.12) for the Butterworth filter and (7.32) for the Chebyshev filter can be written for the inverse Chebyshev filter by using the zero locations from (7.47) and the pole locations from the regular Chebyshev filter given in (7.31) and (7.32). For N even we get

$$F(s) = \frac{\prod_k (s^2 + \omega_{zk}^2)}{\prod_k (s^2 - 2(\sigma_k/(\sigma_k^2 + \omega_k^2))s + 1/(\sigma_k^2 + \omega_k^2))} \quad (7.49)$$

for $k = 1, 3, 5, \ldots, N-1$. For N odd, $F(s)$ has a single pole and therefore is of the form

$$F(s) = \frac{1}{s + 1/\sinh(v_0)} \frac{\prod_k (s^2 + \omega_{zk}^2)}{\prod_k (s^2 - 2(\sigma_k/(\sigma_k^2 + \omega_k^2))s + 1/(\sigma_k^2 + \omega_k^2))} \quad (7.50)$$

for $k = 2, 4, 6, \ldots, N-1$.

Because of the relationships between the locations of the poles of the Butterworth, Chebyshev, and inverse Chebyshev filters, it is easy to write a design program with many common calculations. That is illustrated in Program 9 in the appendix.

7.2.6 Inverse Chebyshev Filter Design Procedures

The natural form for the specifications of an inverse Chebyshev filter is in terms of the response flatness at $\omega = 0$ (to determine the pass band) and a maximum allowable response in the stop band. The filter order and the stop-band ripple are the parameters to be determined by the specifications. The rate of dropoff near the transition from pass band to stop band is similar to the regular Chebyshev filter. Because practical specifications often allow more pass-band ripple than stop-band ripple, the regular Chebyshev filter will usually have a sharper dropoff than the inverse Chebyshev filter will. Under those conditions the inverse Chebyshev filter will have a smoother phase response and less time-domain echo effects.

The stop-band ripple d is simply defined as the maximum value that $|F(j\omega)|$ assumes in the stop band, which is the set of frequencies $1 < \omega < \infty$. An alternative specification is the minimum allowed attenuation over stop band expressed in dB as b. The following formulas relate the stop-band ripple d, the stop-band attenuation b in positive dB, and the transfer function parameter ε in (7.43):

$$\varepsilon = \frac{d}{\sqrt{1 - d^2}} = \frac{1}{\sqrt{10^{b/10} - 1}}, \tag{7.51}$$

$$d = \frac{\varepsilon}{\sqrt{1 + \varepsilon^2}} = 10^{-b/20}, \tag{7.52}$$

$$b = -10 \log\left(\frac{\varepsilon^2}{1 + \varepsilon^2}\right) = -20 \log d. \tag{7.53}$$

In some cases pass-band performance is given not in terms of degree of flatness at $\omega = 0$ but in terms of a minimum allowed magnitude G in the pass band up to a certain frequency ω_p; that is, $1 > |F| > G$ for $0 < \omega < \omega_p < 1$. For a given ε this requirement will determine the order as the smallest positive integer satisfying

$$N \geqslant \frac{\cosh^{-1}\left(\dfrac{G}{\varepsilon\sqrt{1 - G^2}}\right)}{\cosh^{-1}\left(\dfrac{1}{\omega_p}\right)}. \tag{7.54}$$

The design of an inverse Chebyshev filter is summarized in the following steps:

1. The maximum allowed stop-band response must be given in the form of d or b. From this response the parameter ε is calculated by (7.51).
2. The order N is determined from the desired flatness at $\omega = 0$ or from a minimum allowed response for frequencies up to ω_p by (7.54).
3. v_0, $\sinh(v_0)$, and $\cosh(v_0)$ are calculated from (7.29) and (7.31), just as for the regular Chebyshev filter.
4. The pole locations for the prototype Chebyshev filter are calculated from (7.31) or (7.33) and then "inverted" according to (7.48) to give the inverse Chebyshev filter pole locations.
5. The pole locations are combined in (7.48) to give the final filter transfer function denominator.
6. The zero locations are calculated from (7.47) and combined with the pole locations to give the total transfer function (7.49) or (7.50).

Example 7.3. Design of an Inverse Chebyshev Low-Pass Filter
A third-order inverse Chebyshev low-pass filter is desired with a maximum allowed stop-band ripple of $d = 0.1$ or $b = 20$ dB. This value corresponds to an ε of 0.100504 and, together with $N = 3$, results in a v_0 of 0.99774. The scale factors are $\sinh(v_0) = 1.171717$ and $\cosh(v_0) = 1.540429$. The prototype Chebyshev filter transfer function is

$$F(s) = \frac{1}{(s + 1.1717)(s^2 + 1.1717s + 2.0404)}. \tag{7.55}$$

The zeros are calculated from (7.47), and the poles of the prototype are inverted to give, from (7.50), the desired inverse Chebyshev filter transfer function

$$F(s) = \frac{s^2 + \frac{4}{3}}{(s + 0.85345)(s^2 + 0.57425s + 0.490095)}. \tag{7.56}$$

The frequency response of this filter is shown in Fig. 7.12, and the locations of the poles and zeros are shown in Fig. 7.13.

7.2.7 Elliptic Function Filter Properties

In this section a design procedure is developed that uses a Chebyshev error criterion in both the pass band and the stop band. This is the fourth possible combination of Chebyshev and Taylor series approximations in the pass band and the stop band. The resulting filter is called an *elliptic function filter*, because elliptic functions are normally used to calculate the pole and zero locations. It is also sometimes called a *Cauer filter* or a *rational Chebyshev filter*, and it has

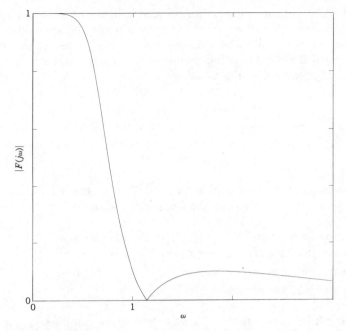

FIGURE 7.12. Example inverse Chebyshev filter frequency response.

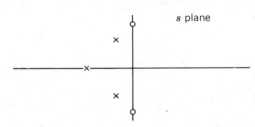

FIGURE 7.13. Pole and zero locations in the s plane for the example inverse Chebyshev filter transfer function.

equal ripple approximation error in the pass band and the stop band.[7,18,19,23]

The error criteria of the elliptic function filter are particularly well suited to the way specifications for filters are often given. Hence, of the four classical filter design methods, elliptic function filter design usually gives the lowest-order filter for a given set of specifications. Unfortunately, the design of this filter is the most complicated. But because it is so efficient, some understanding of the mathematics behind the design is worthwhile.

This section sketches an outline of the theory of elliptic function filter design. One should simply accept the details and properties of the elliptic functions themselves and concentrate on understanding the overall picture. A more complete development is available in references 7, 17 and 23. Straightforward design of elliptic function filters can be accomplished by skipping this section and going directly to Section 7.2.8 and Program 9 in the appendix.

Because the pass-band and stop-band approximations are over the entire bands, a transition band between the two must be defined. Using a normalized pass-band edge, we define the bands by

$$
\begin{aligned}
0 \leqslant \omega \leqslant 1 && \text{(pass band)}, \\
1 < \omega < \omega_s && \text{(transition band)}, \\
\omega_s \leqslant \omega \leqslant \infty && \text{(stop band)}.
\end{aligned}
\qquad (7.57)
$$

See Fig. 7.14.

The characteristics of the elliptic function filter are best described in terms of the four parameters that specify the frequency response:

1. The maximum variation or ripple in the pass band δ_1.
2. The width of the transition band $(\omega_s - 1)$.
3. The maximum response or ripple in the stop band δ_2.
4. The order of the filter N.

The result of the design is that for any three given parameters, the fourth is

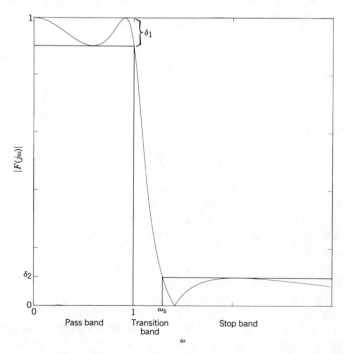

FIGURE 7.14. Elliptic function filter frequency response.

minimum. This description of a filter frequency response is very flexible and powerful.

The form of the frequency-response function is a generalization of that for the Chebyshev filter:

$$\mathcal{F}(j\omega) = |F(j\omega)|^2 = \frac{1}{1 + \varepsilon^2 G^2(\omega)}, \tag{7.58}$$

where

$$\mathcal{F}(s) = F(s)F(-s),$$

with $F(s)$ being the prototype analog filter transfer function similar to (7.2). $G(\omega)$ is a rational function that approximates zero in the pass band and infinity in the stop band. The definition of this function is a generalization of the definition of the Chebyshev polynomial.

Elliptic Functions

To develop analytical expressions for equiripple rational functions, we outline an interesting class of transcendental functions, called the *Jacobian elliptic functions*. These functions can be viewed as a generalization of the normal trigonometric and hyperbolic functions. The elliptic integral of the first kind[21] is defined as

$$u(\phi, k) = \int_0^\phi \frac{dy}{\sqrt{1 - k^2 \sin^2(y)}}. \tag{7.59}$$

The trigonometric sine of the inverse of this function is defined as the Jabocian elliptic sine of u with modulus k and is denoted

$$sn(u, k) = \sin(\phi(u, k)). \tag{7.60}$$

A special evaluation of (7.59) is known as the *complete elliptic integral* $K = u(\pi/2, k)$. It can be shown[21] that $sn(u)$ and most of the other elliptic functions are periodic with periods $4K$ if u is real. Hence, K is also called the *quarter period*. Figure 7.15 is a plot of $sn(u, k)$ for several values of the modulus k. For $k = 0$, $sn(u, 0) = \sin(u)$. As k approaches 1, $sn(u, k)$ looks like a "fat" sine function. For $k = 1$, $sn(u, 1) = \tanh(u)$ and is not periodic.

The quarter period or complete elliptic integral K is a function of the modulus k and is illustrated in Fig. 7.16. For a modulus of zero the quarter period is $K = \pi/2$, and it does not increase much until k approaches 1. It then increases rapidly and goes to infinity as k goes to 1.

Another parameter is the *complementary modulus k'*, defined by

$$k^2 + k'^2 = 1, \tag{7.61}$$

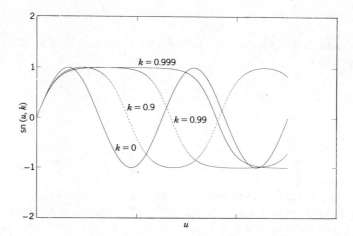

FIGURE 7.15. Elliptic sine of u with moduli k.

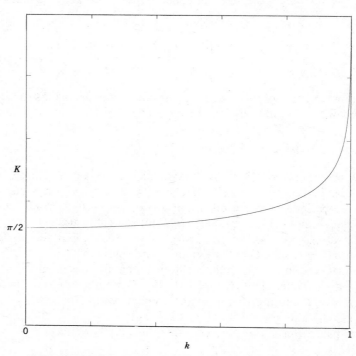

FIGURE 7.16. The complete elliptic integral versus the modulus k.

where we assume $0 < k, k' < 1$; k, k' real. The complete elliptic integral of the complementary modulus is denoted K'.

Other elliptic functions that are rather obvious generalizations are

$$
\begin{aligned}
cn(u, k) &= \cos(\phi(u, k)), \\
sc(u, k) &= \tan(\phi(u, k)), \\
cs(u, k) &= \operatorname{ctn}(\phi(u, k)), \\
nc(u, k) &= \sec(\phi(u, k)), \\
ns(u, k) &= \csc(\phi(u, k)).
\end{aligned}
\tag{7.62}
$$

There are six other elliptic functions that have no trigonometric counterparts.[21] A needed one is

$$
dn(u, k) = \sqrt{1 - k^2 sn^2(u, k)}.
\tag{7.63}
$$

Many interesting properties of the elliptic functions exist.[21] They obey a large set of identities such as

$$
sn^2(u, k) + cn^2(u, k) = 1.
\tag{7.64}
$$

They have derivatives that are elliptic functions. For example,

$$
\frac{d}{du} sn(u, k) = cn(u, k)dn(u, k).
\tag{7.65}
$$

The elliptic functions are the solutions of a set of nonlinear differential equations of the form

$$
\ddot{x} + ax \pm bx^3 = 0.
$$

Some of the most important properties for the elliptic functions are as functions of a complex variable. For a purely imaginary argument

$$
sn(jv, k) = jsc(v, k'), \qquad cn(jv, k) = nc(v, k').
\tag{7.66}
$$

These relations indicate that the elliptic functions, in contrast to the circular and hyperbolic trigonometric functions, are periodic in both the real and imaginary parts of the argument, with periods related to K and K', respectively.

One particular value assumed by the sn function that is important in creating a rational function is

$$
sn(K + jK', k) = \frac{1}{k}.
\tag{7.67}
$$

The Chebyshev Rational Function

The rational function needed in (7.58) is sometimes called a *Chebyshev rational function* because of its equiripple properties. It can be defined in terms of two elliptic functions with moduli k and k_1 by

$$G(\omega) = \text{sn}(n \, \text{sn}^{-1}(\omega, k), k_1).\qquad(7.68)$$

In terms of the intermediate complex variable ϕ, $G(\omega)$ and ω become

$$G(\omega) = \text{sn}(n\phi, k_1), \qquad \omega = \text{sn}(\phi, k).\qquad(7.69)$$

It can be shown[7,17] that $G(\omega)$ is a real-valued rational function if the parameters k, k_1, and n take on special values. Note the similarity of the definition of $G(\omega)$ to the definition of $C_N(\omega)$ in (7.20) and (7.21). In this case, however, n is not necessarily an integer and is not the order of the filter. Requiring that $G(\omega)$ be a rational function requires an alignment of the imaginary periods[7,17] of the two elliptic functions in (7.69). It also requires alignment of an integer multiple of the real periods. The integer multiplier is denoted by N and is the order of the resulting filter.[7,17] These two requirements are stated by the following very important relations:

$$\begin{aligned} nK' &= K'_1 \quad \text{alignment of imaginary periods,} \\ nK &= NK_1 \quad \text{alignment of a multiple of the real periods,} \end{aligned}\qquad(7.70)$$

When the parameter n is removed, (7.70) become

$$\left(\frac{K_1}{K}\right)N = \frac{K'_1}{K'} \quad \text{or} \quad N = \frac{KK'_1}{K'K_1}.\qquad(7.71)$$

These relationships are central to the design of elliptic function filters. N is an odd integer that is the order of the filter. For $N = 5$ the resulting rational function is shown in Fig. 7.17.

This function is the basis of the approximation necessary for the optimal filter frequency response. It approximates zero over the frequency range $-1 < \omega < 1$ by an equiripple oscillation between $+1$ and -1. It also approximates infinity over the range $1/k < |\omega| < \infty$ by a reciprocal oscillation that keeps $|F(\omega)| > 1/k_1$. The zero approximation is normalized in both the frequency range and the $F(\omega)$ values to unity. The infinity approximation has its frequency range set by the choice of the modulus k, and the minimum value of $|F(\omega)|$ is set by the choice of the second modulus k_1.

If k and k_1 are determined from the filter specifications, they in turn determine the complementary moduli k' and k'_1, which altogether determine the four values of the complete elliptic integral K needed to determine the order N in (7.71). In general, this sequence of events will not produce an integer. In practice,

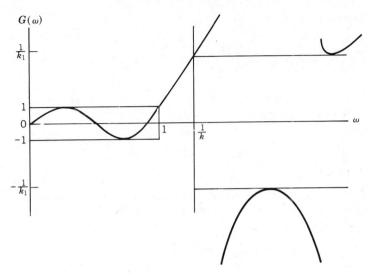

FIGURE 7.17. Fifth-order elliptic rational function.

however, the next larger integer is used, and either k or k_1 (or perhaps both) is altered to satisfy (7.71).

In addition to the two-band equiripple characteristics, $G(\omega)$ has another interesting and valuable property. The pole and zero locations have a reciprocal relationship[26,56] that can be expressed by

$$G(\omega)G\left(\frac{\omega_s}{\omega}\right) = \frac{1}{k_1} \qquad \text{where} \quad \omega_s = \frac{1}{k}.$$

This property states that if the zeros of $G(\omega)$ are located at ω_{zi}, the poles are located at

$$\omega_{pi} = \frac{1}{k\omega_{zi}}. \tag{7.72}$$

If the zeros are known, the poles are known, and vice versa. A similar relation exists between the points of zero derivatives in the 0 to 1 region and those in the $1/k$ to ∞ region.

The zeros of $G(\omega)$ are found from (7.69) by requiring

$$G(\omega) = \text{sn}(n\phi, k_1) = 0,$$

which implies

$$n\phi = 2K_1 i \qquad \text{for } i = 0, 1, \dots.$$

From (7.69) this relation gives

$$\omega_{zi} = \text{sn}\left(\frac{2K_1 i}{n}, k\right), \qquad i = 0, 1, \ldots. \tag{7.73}$$

Using (7.70), we can reformulate (7.73) so that n and K_1 are not needed. For N odd the zero locations are

$$\omega_{zi} = \text{sn}\left(\frac{2Ki}{N}, k\right), \qquad i = 0, 1, \ldots. \tag{7.74}$$

We find the pole locations from these zero locations by using (7.72). The locations of the zero-derivative points are given by

$$\omega_{di} = \text{sn}\left[\frac{K(2i + 1)}{N}, k\right] \tag{7.75}$$

in the 0 to 1 region, and the corresponding points in the $1/k$ to ∞ region are found from (7.72).

These relations assume that N is an odd integer. A modification for even N is necessary. For proper alignment of the real periods the original definition of $G(\omega)$ in (7.69) is changed to

$$G(\omega) = \text{sn}(n\phi + K_1, k_1), \tag{7.76}$$

which, for N even, gives for the zero locations

$$\omega_{zi} = \text{sn}\left[\frac{(2i + 1)K_1}{n}, k\right]. \tag{7.77}$$

The even and odd N cases of (7.74) and (7.77) can be combined to give

$$\omega_{zi} = \pm\text{sn}\left(\frac{iK}{N}, k\right) \tag{7.78}$$

for

$$i = \begin{cases} 0, 2, 4, \ldots, N - 1 & \text{for } N \text{ odd,} \\ 1, 3, 5, \ldots, N - 1 & \text{for } N \text{ even,} \end{cases}$$

with the poles determined from (7.72).

Note that it is possible to determine $G(\omega)$ from k and N without explicitly using k_1 or n. Values for k_1 and n are implied by the requirements of (7.70) or (7.71).

Zero Locations
The locations of the zeros of the filter transfer function $F(\omega)$ are easily found, since they are the same as the poles of $G(\omega)$, given in (7.78).

$$\omega_{zi} = \frac{\pm 1}{k \, \text{sn}(iK/N, k)} \tag{7.79}$$

for

$$i = \begin{cases} 0, 2, 4, \ldots, N-1 & N \text{ odd,} \\ 1, 3, 5, \ldots, N-1 & N \text{ even.} \end{cases}$$

These zeros are purely imaginary and lie on the ω axis.

Pole Locations
The pole locations are somewhat more complicated to find. We use an approach similar to that used for the Chebyshev filter in (7.25). $\mathscr{F}(s)$ becomes infinite when

$$1 + \varepsilon^2 G^2 = 0$$

or

$$G = \pm j\left(\frac{1}{\varepsilon}\right). \tag{7.80}$$

Using (7.69) and the periodicity of sn(u, k), we then get

$$G = \text{sn}(n\phi + 2K_1 i, k_1) = \pm j\left(\frac{1}{\varepsilon}\right)$$

or

$$\phi = \frac{-2K_1 i + \text{sn}^{-1}[j(1/\varepsilon), k_1]}{n} \tag{7.81}$$

Define v_0 to be the second term in (7.81) by

$$jv_0 = \frac{\text{sn}^{-1}[j(1/\varepsilon), k_1]}{n},$$

which is similar to (7.29) for the Chebyshev case. Using the properties of sn of an imaginary variable and (7.71), we obtain

$$v_0 = \left(\frac{K}{NK_1}\right)\text{sc}^{-1}\left(\frac{1}{\varepsilon}, k_1'\right). \tag{7.82}$$

The poles are now found from (7.69), (7.81), and (7.82) to be

$$s_{pi} = j\,\text{sn}\!\left(\frac{Ki}{N} + jv_0,\, k\right).$$

(7.82)

This equation can be more clearly written by using the summation formula[21] for the elliptic sine function:

$$s_{pi} = \frac{\text{cn dn sn}'\,\text{cn}' + j\,\text{sn dn}'}{1 - \text{dn}^2\,\text{sn}'^2},$$

(7.84)

where

$$\text{sn} = \text{sn}\!\left(\frac{Ki}{N},\, k\right), \qquad \text{cn} = \text{cn}\!\left(\frac{Ki}{N},\, k\right), \qquad \text{dn} = \text{dn}\!\left(\frac{Ki}{N},\, k\right),$$
$$\text{sn}' = \text{sn}(v_0,\, k'), \qquad \text{cn}' = \text{cn}(v_0,\, k'), \qquad \text{dn}' = \text{dn}(v_0,\, k'),$$

(7.85)

for

$$N \text{ odd:} \qquad i = 0,\, 2,\, 4,\, \ldots,$$
$$N \text{ even:} \qquad i = 1,\, 3,\, 5,\, \ldots.$$

   The theory of Jacobian elliptic functions can be found in references 17 and 21 and its application to filter design in references 7, 17–19, and 23. The best techniques for calculating the elliptic functions seem to use the arithmetic-geometric mean; efficient algorithms are presented in reference 22. A design program is given in reference 23 and the versatile FORTRAN Program 9, which is easily related to the theory in this chapter, is given in the appendix.
   The transfer function $F(s)$ pole locations can also be found by obtaining the zeros from (7.79) and finding $G(\omega)$ by using the reciprocal relation of the poles and zeros (7.72). $F(s)$ is constructed from $G(\omega)$, ε from (7.58), and the poles are found by a root-finding algorithm. Another possibility is to find the zeros from (7.79) and the poles from the methods for finding a Chebyshev pass band from arbitrary zeros.[17,24,25] These approaches avoid calculating v_0 by (7.82) or determining k from K/K', as is described in reference 23. The efficient algorithms for evaluating the elliptic functions and the common use of powerful computers make these alternatives less attractive now.

Summary

This section outlined the basic properties of the Jacobian elliptic functions and gave the necessary conditions for an equiripple rational function to be defined in terms of them. This rational function was then used to construct a filter transfer function with equiripple properties. Formulas were derived to calculate the pole and zero locations for the filter transfer functions and to relate design specifications to the functions. These formulas require the evaluation of elliptic functions and are implemented in Program 9.

7.2.8 Elliptic Function Filter Design Procedures

The equiripple rational function $G(\omega)$ is used to describe an optimal frequency-response function $F(j\omega)$ and to design the corresponding filter. The squared magnitude frequency-response function from (7.58) is

$$|F(j\omega)|^2 = \frac{1}{1 + \varepsilon^2 G^2(\omega)},\qquad (7.86)$$

with $G(\omega)$ defined by (7.68) and (7.76), and ε a parameter that controls the pass-band ripple. The plot of this function for $N = 5$ illustrates the relation to the various specification parameters. Figure 7.18 shows that the pass-band ripple is measured by δ_1, the stop-band ripple by δ_2, and the normalized transition band by ω_s. The previous section showed in (7.72) that

$$\omega_s = \frac{1}{k},$$

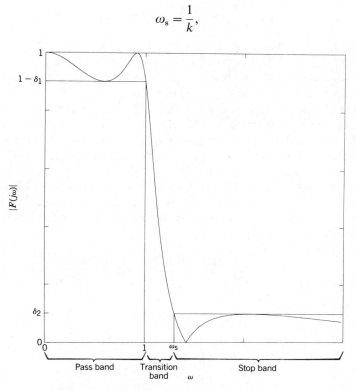

FIGURE 7.18. Elliptic-function filter frequency response.

which means that the width of the transition band determines k. Remember that in this developement we have assumed a pass-band edge normalized to unity. For the unnormalized case the pass-band edge is ω_p, and the stop-band edge becomes

$$\omega_s = \frac{\omega_p}{k}. \tag{7.87}$$

The stop-band performance is described in terms of the ripple δ_2 normalized to a maximum pass-band response of unity or in terms of the attenuation b in the stop band expressed in positive dB, if we assume a maximum pass-band response of zero dB. The stop-band ripple and attenuation are determined from (7.86) and Fig. 7.18 to be

$$\delta_2^2 = 10^{-b/10} = \frac{1}{1 + \varepsilon^2/k_1^2}. \tag{7.88}$$

Rearranging gives k_1 in terms of the stop-band ripple or attenuation

$$k_1^2 = \frac{\varepsilon^2}{1/\delta_2^2 - 1} = \frac{\varepsilon^2}{10^{b/10} - 1}. \tag{7.89}$$

The order N of the filter depends on k and k_1, as shown in (7.71). Equations (7.87), (7.89), and (7.71) determine the relation of the frequency-response specifications and the elliptic function parameters. The location of the transfer function poles and zeros must then be determined.

Because of the required relationships of (7.71) and because the order N must be an integer, the pass-band ripple, stop-band ripple, and transition band cannot be independently set. Several straightforward procedures can be used that will always meet two of the specifications and exceed the third.

The first design step is generally the determination of the order N from the desired pass-band ripple δ_1, the stop-band ripple δ_2, and the transition band controlled by ω_s. The following formulas determine the moduli k and k_1 from the pass-band ripple δ_1 or its dB equivalent a and the stop-band ripple δ_2 or its dB attenuation equivalent b:

$$\varepsilon = \sqrt{\frac{2\delta_1 - \delta_1^2}{1 - 2\delta_1 + \delta_1^2}} = \sqrt{10^{a/10} - 1}, \tag{7.90}$$

$$k_1 = \frac{\varepsilon}{\sqrt{1/\delta_2^2 - 1}} = \frac{\varepsilon}{\sqrt{10^{b/10} - 1}}, \tag{7.91}$$

$$k_1' = \sqrt{1 - k_1^2},$$

$$k = \frac{\omega_p}{\omega_s}, \qquad k' = \sqrt{1 - k^2}. \tag{7.92}$$

The order N is the smallest integer satisfying

$$N \geqslant \frac{KK'_2}{K'K_1}. \tag{7.93}$$

This integer order N will not, in general, exactly satisfy (7.71)—that is, it will not satisfy (7.93) with equality. Either k or k_1 must to recalculated to satisfy (7.71) and (7.93). The various possibilities for this are developed here.

Methods for Meeting Specifications

A. Fixed-Order, Pass-Band Ripple, and Transition Band
Given N from (7.93) and the specifications δ_1, ω_p, and ω_s, we find the parameters ε and k from (7.90) and (7.92). From k the complete elliptic integrals K and K' are calculated.[22] From (7.71) the ratio K/K' determines the ratio K'_1/K_1. Using numerical methods from references 21 and 23, we calculate k_1, which gives the desired δ_1, ω_p, and ω_s and minimizes the stop-band ripple δ_2 (or maximizes the stop-band attenuation b).

Using these parameters, we calculate the zeros from (7.79) and the poles from (7.84). Note that the zero locations depend not on ε or k_1 but only on N and ω_s. This dependence makes the tradeoff between stop band and pass band occur in (7.91) and only affects the calculation of v_0 in (7.82).

This approach, which minimizes the stop-band ripple, is used in the IIR filter design program in the appendix.

B. Fixed-Order, Stop-Band Rejection, and Transition Band
Given N from (7.93) and the specifications δ_2, ω_p, and ω_s, we find the parameter k from (7.92). From k the complete elliptic integrals K and K' are calculated[22]. From (7.71) the ratio K/K' determines the ratio K'_1/K_1. Using numerical methods from references 21 and 23, we calculate k_1. From k_1 and δ_2, ε and δ_1 are found from

$$\varepsilon = k_1 \sqrt{\frac{1}{\delta_2^2} - 1} \tag{7.94}$$

and

$$\delta_1 = 1 - \frac{1}{\sqrt{1 + \varepsilon^2}}. \tag{7.95}$$

This set of parameters gives the desired ω_p, ω_s, and stop-band ripple and minimizes the pass-band ripple. The zero and pole locations are found as in A.

C. Fixed-Order, Stop-Band, and Pass-Band Ripple
Given N from (7.93) and the specifications δ_1, δ_2, and either ω_p or ω_s, we find the

parameters ε and k_1 from (7.90) and (7.91). From k_1 the complete elliptic integrals K_1 and K'_1 are calculated[22]. From (7.71) the ratio K_1/K'_1 determines the ratio K'/K. Using numerical methods from references 21 and 23, we calculate k which gives the desired pass-band and stop-band ripple and minimizes the transition band width. The pole and zero locations are found as before.

D. An Approximation

After the order N is found from (7.93), in many filter design programs the design proceeds with the original ε, k, and k_1, even though they do not satisfy (7.71). The resulting design has the desired transition band, but pass-band and stop-band ripple are smaller than specified. This procedure avoids calculating the modulus k or k_1 from a ratio of complete elliptic integrals, which was necessary in all three cases before, but produces results that are difficult to predict exactly.

Example 7.4. Design of a Third-Order Elliptic Function Low-Pass Filter

A low-pass elliptic function filter is desired with a maximum pass-band ripple of $\delta_1 = 0.1$ or $a = 0.91515$ dB, a maximum stop-band ripple of $\delta_2 = 0.1$ or $b = 20$ dB rejection, and a normalized stop-band edge of $\omega_s = 1.3$ rad/s. The first step is to determine the order of the filter.

From ω_s we calculate the modulus k and then, using the relations in (7.92), the complementary modulus. Special numerical algorithms illustrated in Program 9 are then used to find the complete elliptic integrals K and K'.[22]

$$k = \frac{1}{1.3} = 0.769231, \qquad k' = \sqrt{1 - k^2} = 0.638971,$$
$$K = 1.940714, \qquad\qquad K' = 1.783308.$$

From δ_1, we calculate ε, using (7.90), and from ε and δ_2, we calculate k_1 from (7.91). We then determine k'_1, K_1, and K'_1:

$$\varepsilon = 0.4843221 \qquad \text{as for the Chebyshev example,}$$
$$k_1 = 0.0486762, \qquad k'_1 = 0.9988146,$$
$$K_1 = 1.571727, \qquad K'_2 = 4.4108715.$$

The order is obtained from (7.71) by calculating

$$\frac{KK'_1}{K'K_1} = 3.0541,$$

which is close enough to 3 to set $N = 3$. Rather than recalculate k and k_1, we use the already calculated values, as discussed in design method D. We find the zeros from (7.79), using only N and k determined earlier.

$$\omega_z = \frac{\pm 1}{k \, \mathrm{sn}(2K/N, \, k)} = \pm 1.430207.$$

Finding the pole locations requires calculating v_0 from (7.82), which is somewhat complicated. It is carried out with the algorithms in Program 9.

$$v_0 = \frac{K}{NK_1} sc^{-1}\left(\frac{1}{\varepsilon}, k_1'\right) = 0.6059485.$$

Using the values of v_0, k, and N, we calculate the elliptic functions in (7.85):

$$sn' = .557986, \quad cn' = 0.829850, \quad dn' = 0.934281,$$

which for the single, real pole corresponding to $i = 0$ in (7.84) gives

$$s_p = 0.672393.$$

For the complex-conjugate pair of poles corresponding to $i = 2$, the other elliptic functions in (7.85) are

$$sn = 0.908959, \quad cn = 0.416886, \quad dn = 0.714927,$$

which gives, from (7.84),

$$s_p = 0.164126 \pm j1.009942$$

for the poles. The complete transfer function is

$$F(s) = \frac{s^2 + 2.045492}{(s + 0.672393)(s^2 + 0.328252s + 1.046920)}. \tag{7.96}$$

The frequency response of this filter is in Fig. 7.19, and the locations of the poles and zeros are in Fig. 7.20. This design should be compared to the Chebyshev and inverse Chebyshev designs.

7.2.9 Optimality of the Four Classical Filter Designs

It is important in filter design to choose the appropriate type. Since the filters are optimal in all cases, it is necessary to understand in what sense they are optimal.

The classical Butterworth filter is optimal in the sense that it is the best Taylor series approximation to the ideal low-pass filter magnitude at both $\omega = 0$ and $\omega = \infty$.

The Chebyshev filter gives the smallest maximum magnitude error over the entire pass band of any filter that is also a Taylor series approximation at $\omega = \infty$ to the ideal magnitude characteristic.

The inverse Chebyshev filter is a Taylor series approximation to the ideal magnitude response at $\omega = 0$ and minimizes the maximum error in the approximation to zero over the stop band. Or we can say it maximizes the minimum rejection of the filter over the stop band.

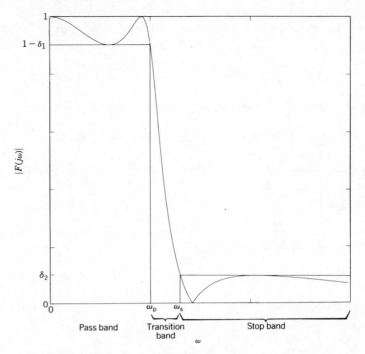

FIGURE 7.19. Example elliptic function filter frequency response.

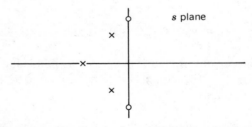

FIGURE 7.20. Pole and zero locations in the *s* plane for the example elliptic function filter.

The elliptic function filter (Cauer filter) considers the four parameters of the filter: the pass-band ripple, the transition band width, the stop-band ripple, and the order of the filter. For given values of any three of the four, the fourth is minimized.

Remember that all four of these filter designs are magnitude approximations and do not address the phase frequency response or the time-domain character-istics. For most designs the Butterworth filter has the smoothest phase curve, followed by the inverse Chebyshev filter, Chebyshev filter, and elliptic function filter.

Recall that in addition to the four filters described in this section, the more general Taylor series method described in (7.8) allows arbitrary zero locations to be specified but retains the optimal character at $\omega = 0$. A design similar to this can be obtained by replacing ω by $1/\omega$, which allows setting $|F(\omega)|^2$ equal to unity at arbitrary frequencies in the pass band and having a Taylor series approximation to zero at $\omega = \infty$. Similar modifications of the Chebyshev filters are covered in references 24 and 25.

These basic normalized low-pass filters can have the pass-band edge moved from unity to any desired value by a simple change of frequency variable: $k\omega$ replace with ω. They can be converted to high-pass, bandpass, or band-rejection filters by various changes, such as replacing ω by k/ω or by $a\omega + b/\omega$. In all of these cases the optimality is maintained, because the basic low-pass approximation is to a piecewise constant ideal. An approximation to a nonpiecewise constant ideal, such as a differentiator, may not be optimal after a frequency change of variables.

In some cases, especially where time-domain characteristics are important, ripples in the frequency response cause irregularities, such as echoes in the time response. For that reason the Butterworth and Chebyshev II filters are more desirable than their frequency response along might indicate. A fifth approximation has been developed[20] that is similar to the Butterworth. It requires not a Taylor series approximation at $\omega = 0$ but only that the response monotonically decrease in the pass band, thus giving a narrower transition region than the Butterworth but without the ripples of the Chebyshev.

7.2.10 Frequency Transformations

In addition to the low-pass frequency response, other basic ideal responses are often needed in practice. The ideal high-pass filter rejects signals with frequencies below a certain value and passes those with frequencies above that value. The ideal bandpass filter passes only a band of frequencies, and the ideal band-rejection filter completely rejects a band of frequencies. These ideal frequency responses are illustrated in Fig. 7.21.

This section presents a method for designing the three new filters by using a frequency transformation on the basic low-pass design. When used on the four approximations covered in Sections 7.2.1 through 7.2.8, they preserve optimality. This procedure is used in the FREQXFM( ) subroutine of Program 9.

The High-Pass Filter
The frequency response illustrated in Fig. 7.21*b* can be obtained from that in 7.21*a* by replacing the complex frequency variable s in the transfer function by $1/s$. This change of variable maps zero frequency to infinity, maps unity into unity, and maps infinity to zero. It turns the complex s plane inside out and leaves the unit circle alone.

In the design procedure the desired band edge ω_0 for the high-pass filter is mapped by $1/\omega_0$ to give the band edge for the prototype low-pass filter. This low-pass filter is next designed by one of the optimal procedures already covered

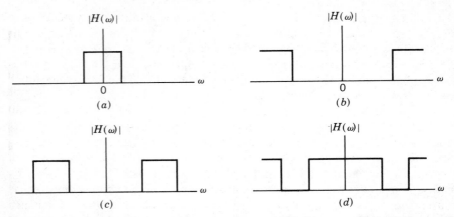

FIGURE 7.21. The four basic ideal frequency responses. (a) Ideal low pass; (b) ideal high pass; (c) ideal bandpass; (d) ideal band rejection.

and then converted to a high-pass transfer function by replacing s by $1/s$. If an elliptic function filter approximation is used, both the pass-band edge ω_p and the stop-band edge ω_s are transformed. Because most optimal low-pass design procedures give the designed transfer function in factored form from explicit formulas for the poles and zeros, the transformation can be performed on each pole and zero to give the high-pass transfer function in factored form.

The Bandpass Filter
To convert the low-pass filter of Fig. 7.21a into that of 7.21c, we need a more complicated frequency transformation. To reduce confusion, we denote the complex frequency variable for the prototype analog filter transfer function by p and that for the transformed analog filter by s. The transformation is given by

$$p = \frac{s^2 + \omega_0^2}{s}.$$

This change of variables doubles the order of the filter, maps the origin of the s plane to $\pm j\omega_0$, and maps $\pm\infty$ to 0 and ∞. The entire ω axis of the prototype response is mapped between 0 and $+\infty$ on the transformed responses. It is also mapped onto the left-half-axis between $-\infty$ and 0. See Fig. 7.22. For the transformation to give $-\omega_p = (\omega_2^2 - \omega_0^2)/\omega_2$ and $\omega_p = (\omega_3^2 - \omega_0^2)/\omega_3$, the "center" frequency ω_0 must be

$$\omega_0 = \sqrt{\omega_2\omega_3}. \tag{7.97}$$

However, because $-\omega_s = (\omega_1^2 - \omega_0^2)\omega_1$ and $\omega_s = (\omega_4^2 - \omega_0^2)/\omega_4$, the center frequency must also be

$$\omega_0 = \sqrt{\omega_1\omega_4}.$$

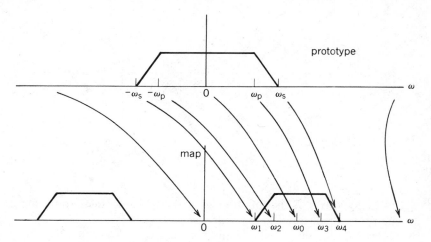

FIGURE 7.22. Low-pass to bandpass frequency transformation.

This means that only three of the four band-edge frequencies (ω_1, ω_2, ω_3, and ω_4) can be independently specified. Normally, ω_0 is determined by ω_2 and ω_3, which then specify the prototype pass-band edge by

$$\omega_p = \frac{\omega_3^2 - \omega_0^2}{\omega_3}. \qquad (7.98)$$

Using the same ω_0, we set the stop-band edge by either ω_1 or ω_4, whichever gives the smaller ω_s.

$$\omega_s = \frac{\omega_4^2 - \omega_0^2}{\omega_4} \quad \text{or} \quad \frac{\omega_0^2 - \omega_1^2}{\omega_1}.$$

The finally designed bandpass filter will meet both pass-band edges and one transition bandwidth, but the other will be narrower than originally specified. That is not a problem with the Butterworth or either of the Chebyshev approximations because they have either pass-band edges or stop-band edges. The elliptic function has both.

After we calculate the band edges for the prototype low-pass filter ω_p and/or ω_s, we design the filter by one of the optimal approximation methods discussed in this section or by any other means. Because most of these methods give the pole and zero locations directly, they can be individually transformed to give the bandpass filter transfer function in factored form. It is accomplished by solving $s^2 - ps + \omega_0^2$ from the original transformation to give

$$s = \frac{p \pm \sqrt{p^2 - 4\omega_0^2}}{2} \qquad (7.99)$$

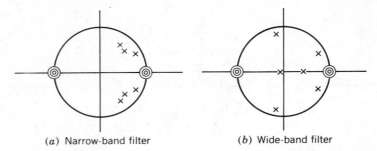

(a) Narrow-band filter   (b) Wide-band filter

FIGURE 7.23. Pole and zero locations for a bandpass filter. (a) Narrow-band filter; (b) Wide-band filter.

for the root locations. This equation gives two transformed roots for each prototype root, which doubles the order as expected.

Examples of two third-order, bandpass, Chebyshev filter pole and zero locations (after converting to a digital filter) on the z plane are shown in Fig. 7.23. The roots that result from transforming the real pole of an odd-order prototype cause some complication in programming this procedure. Program 9 should be studied to understand how this is carried out.

The Band-Reject Filter

To design a filter that will reject a band of frequencies, we use a frequency transformation of the form

$$p = \frac{s}{s^2 + \omega_0^2}$$

on the prototype low-pass filter. This transforms the origin of the p plane into both the origin and infinity of the s plane. It maps infinity in the p plane into $j\omega_0$ in the s plane. See Fig. 7.24.

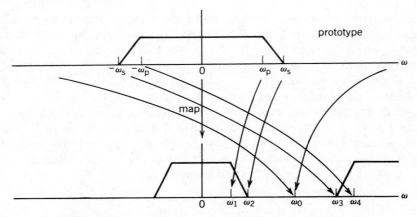

FIGURE 7.24. Low-Pass to band-rejection frequency transformation.

Similar to the bandpass case, the transformation must give $-\omega_p = \omega_4/(\omega_0^2 - \omega_4^2)$ and $\omega_p = \omega_1(\omega_0^2 - \omega_1^2)$. A similar relation of ω_s to ω_2 and ω_3 requires that the center frequency ω_0 must be

$$\omega_0 = \sqrt{\omega_1\omega_4} = \sqrt{\omega_2\omega_3}.$$

As before, only three of the four band-edge frequencies can be independently specified. Normally, ω_0 is determined by ω_1 and ω_4, which then specify the prototype pass-band edge by

$$\omega_p = \frac{\omega_1}{\omega_0^2 - \omega_1^2}. \tag{7.100}$$

Using the same ω_0, we set the stop-band edge by either ω_2 or ω_3, whichever gives the smaller ω_s.

$$\omega_s = \frac{\omega_2}{\omega_0^2 - \omega_2^2} \quad \text{or} \quad \frac{\omega_3}{\omega_3^2 - \omega_0^2}.$$

The finally designed bandpass filter will meet both pass-band edges and one transition bandwidth, but the other will be narrower than originally specified. This occurs not with the Butterworth or either Chebyshev approximation but only with the elliptic function.

After we calculate the band edges for the prototype low-pass filter ω_p and/or ω_s, we design the filter. The poles and zeros of this filter are individually transformed to give the band-rejection filter transfer function in factored form. It is carried out by solving $s^2 - (1/p)s + \omega_0^2$ to give for the root locations

$$s = \frac{1/p \pm \sqrt{(1/p)^2 - 4\omega_0^2}}{2}. \tag{7.101}$$

Examples of two third-order band-reject Chebyshev filter pole and zero locations (after converting to a digital filter) on the z plane are shown in Fig. 7.25.

A more complicated set of transformations could be developed by using a general map of $s = f(s)$ with a higher order. Several pass bands or stop bands could be specified, but the calculations become fairly complicated.

Although this method of transformation is a powerful and simple way for designing bandpass and band-reject filters, it does impose certain restrictions. A Chebyshev bandpass filter will be equiripple in the pass band and maximally flat at both zero and infinity, but the transformation forces the degree of flatness at zero and infinity to be equal. The elliptic function bandpass filter will have the same number of ripples in both stop bands even if they are of very different widths. These restrictions are usually considered mild when compared with the complexity of alternative design methods.

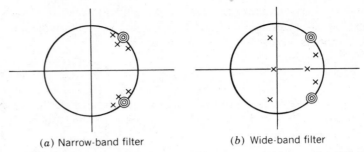

(*a*) Narrow-band filter (*b*) Wide-band filter

FIGURE 7.25. Pole and zero locations for a band rejection filter. (*a*) Narrow-band filter; (*b*) wideband filter.

7.3. CONVERSION OF ANALOG-TO-DIGITAL TRANSFER FUNCTIONS

For mathematical convenience the four classical IIR filter transfer functions were developed in Section 7.2 in terms of the Laplace transform rather than the z transform. The prototype Laplace transform transfer functions are descriptions of analog filters. In this section they are converted to z transform transfer functions for implementation as IIR digital filters.

Of the several methods described over the history of digital filters for converting analog systems to digital systems, two have proven to be useful for most applications. The first is the *impulse-invariant method*, which gives a digital filter with an impulse response exactly equal to samples of the prototype analog filter. The second method uses a frequency mapping to convert the analog filter to a digital filter. It has the desirable property of preserving the optimality of the four classical approximations developed in the last section. This section develops the theory and design formulas to implement both of these conversion approaches.

7.3.1 The Impulse-Invariant Method

Although the transfer functions in Section 7.2 were designed with criteria in the frequency domain, the impulse-invariant method converts them into digital transfer functions by a time-domain constraint.[1–3] The digital filter designed by the impulse-invariant method is required to have an impulse response exactly equal to equally spaced samples of the impulse response of the prototype analog filter. If the analog filter has a transfer function $F(s)$ with an impulse response $f(t)$, the impulse response of the digital filter $h(n)$ is required to match the samples of $f(t)$. For samples at intervals of T seconds the impulse response is

$$h(n) = f(t)\Big|_{t=Tn} = f(Tn). \tag{7.102}$$

The transfer function of the digital filter is the z transform of the impulse response of the filter,

$$H(z) = \sum_{n=0}^{\infty} h(n)z^{-n}.$$

The transfer function of the prototype analog filter is always a rational function,

$$F(s) = \frac{B(s)}{A(s)},$$

where $B(s)$ is the numerator polynomial with roots that are the zeros of $F(s)$, and $A(s)$ is the denominator with roots that are the poles of $F(s)$. If $F(s)$ is expanded in terms of partial fractions, it can be written as

$$F(s) = \sum_{i=1}^{N} \frac{K_i}{s + s_i}. \tag{7.103}$$

The impulse response of this filter is the inverse Laplace transform of (7.103), which is

$$f(t) = \sum_{i=1}^{N} K_i e^{-s_i t}.$$

Sampling this impulse response every T seconds gives

$$f(nT) = \sum_{i=1}^{N} K_i e^{-s_i n T} = \sum_{i=1}^{N} K_i (e^{-s_i T})^n.$$

The basic requirement of (7.102) gives

$$H(z) = \sum_{n=0}^{\infty} \left[\sum_{i=1}^{N} K_i (e^{-s_i T})^n \right] z^{-n},$$

$$= \sum_{i=1}^{N} \frac{K_i z}{z - e^{-s_i T}}, \tag{7.104}$$

which is clearly a rational function of z and is the transfer function of the digital filter, which has samples of the prototype analog filter as its impulse response.

This method has its requirements set in the time domain, but the frequency response is important. In most cases the prototype analog filter is one of the classical types from Section 7.2, which is optimal in the frequency domain. If the frequency response of the analog filter is denoted by $F(j\omega)$ and the frequency response of the digital filter designed by the impulse-invariant method is $H(\omega)$, it can be shown, in a development similar to that used for the sampling theorem, that

$$H(\omega) = \frac{1}{T} \sum_{k=-\infty}^{\infty} F\left(j\left(\omega - \frac{2\pi k}{T}\right)\right). \tag{7.105}$$

The frequency response of the digital filter is a periodically repeated version of the frequency response of the analog filter. This produces an overlapping of the analog response. Thus optimality is not preserved in the same sense in which the analog prototype was optimal. It is a similar phenomenon to the aliasing that occurs when sampling a continuous-time signal to obtain a digital signal in A/D conversion. If $F(j\omega)$ is an analog low-pass filter that goes to zero as ω goes to infinity, the effects of the folding can be made small by high sampling rates (small T).

The impulse-invariant design method can be summarized in the following steps:

1. Design a prototype analog filter with transfer function $F(j\omega)$.
2. Make a partial fraction expansion of $F(j\omega)$ to obtain the N values for K_i and s_i for (7.100).
3. Form the digital transfer function $H(z)$ from (7.104) to give the desired design.

The characteristics of the designed filter are the following:

1. It has N poles, the same as the analog filter.
2. It is stable if the analog filter was stable. We see this from the change of variables in the denominator of (6.70), which maps the left-half of the s plane inside the unit circle in the z plane.
3. The frequency response is a folded version of the analog filter, and the optimal properties of the analog filter are not preserved.
4. The cascade of two impulse-invariant designed filters is not impulse-invariant with the cascade of the two analog prototypes. In other words, the filter must be designed in one step.

This method is sometimes used to design digital filters, but because the relation between the analog and digital systems is specified in the time domain, it is more useful in designing a digital simulation of an analog system. Unfortunately, the properties of this class of filters depend on the input. If a filter is designed so that its impulse response is the sampled impulse response of the analog filter, its step response will not be the sampled step response of the analog filter.

A step-invariant filter can be designed by first multiplying the analog filter transfer function $F(s)$ by $1/s$, which is the Laplace transform of a step function.

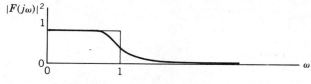

(a) The Butterworth Lowpass Prototype Analog Filter

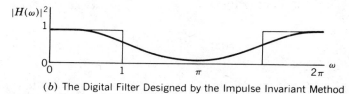

(b) The Digital Filter Designed by the Impulse Invariant Method

FIGURE 7.26. Impulse-invariant design of a Butterworth filter. (a) Butterworth low-pass prototype analog filter; (b) digital filter designed by the impulse-invariant method.

This product is then expanded in partial fractions just as $F(s)$ was in (7.100), and the same substitution is made as in (7.104), giving a z transform. After the z transform of a step is removed, the digital filter has the step-invariant property. This idea can be extended to other input functions, but the impulse-invariant version is the most common.

Another modification to the impulse-invariant method is known as the matched z transform, covered in reference 1, but it is less useful.

An example of a Butterworth low-pass filter used to design a digital filter by the impulse-invariant method is shown in Fig. 7.26. Note that the frequency response does not go to zero at the highest frequency $\omega = \pi$. We can make it as small as we like by increasing the sampling rate, but this is more expensive to implement. Because the frequency response of the prototype analog filter for an inverse Chebyshev or elliptic function filter does not necessarily go to zero as ω goes to infinity, the effects of folding on the digital frequency response are poor. No amount of sampling rate increase will change that. The same problem exists for a high-pass filter. Therefore care must be exercised in using the impulse-invariant design method.

7.3.2 The Bilinear Transformation

A second method for converting an analog prototype filter into a desired digital filter is the *bilinear transformation*. This method is entirely a frequency-domain method, and, as a result, some of the optimal properties of the analog filter are preserved. As was the case with the impulse-invariant method, the time interval is not normalized to unity but is explicitly denoted by the sampling interval T with units of seconds. The bilinear transformation is a change of variables (a mapping) that is linear in the numerator and the denominator.[1-4] The usual form is

$$s = \frac{2}{T}\frac{z-1}{z+1}. \qquad (7.106)$$

The z transform transfer function of the digital filter $H(z)$ is obtained from the Laplace transform transfer function $F(s)$ of the prototype filter by substituting for s the bilinear form of (7.106).

$$H(z) = F\left(\frac{2}{T}\frac{z-1}{z+1}\right). \qquad (7.107)$$

This operation can be reversed by solving (7.106) for z and substituting this into $H(z)$ to obtain $F(s)$. This reverse operation is also a bilinear transformation of the form

$$z = \frac{2/T + s}{2/T - s}. \qquad (7.108)$$

To consider the frequency response, we evaluate the Laplace variable s on the imaginary axis and the z transform variable z on the unit circle. We set

$$s = ju \quad \text{and} \quad z = e^{j\omega T}, \qquad (7.109)$$

which gives the relation of the analog frequency variable u to the digital frequency variable ω, from (7.109) and (7.106), as

$$u = \frac{2}{T}\tan\left(\frac{\omega T}{2}\right). \qquad (7.110)$$

The bilinear transform maps the infinite imaginary axis in the s plane onto the unit circle in the z plane. It maps the infinite interval of $-\infty < u < \infty$ of the analog frequency axis onto the finite interval $-\pi/2 < \omega < \pi/2$ of the digital frequency axis. See Figs. 7.27 and 7.28. There is no folding or aliasing of the prototype frequency response, but there is a compression of the frequency axis that becomes extreme at high frequencies. This compression is shown in Fig. 7.28 from the relation (7.110). Near zero frequency, the relation of u and ω is essentially linear. The compression increases as the digital frequency ω nears $\pi/2$. This nonlinear compression is called *frequency warping*. The conversion of $F(s)$ to $H(z)$ with the bilinear transformation does not change the values of the frequency response, but it changes the frequencies where the values occur.

In the design of a digital filter the effects of the frequency warping must be taken into account. The prototype filter frequency scale must be prewarped so that after the bilinear transform the critical frequencies are in the correct places.

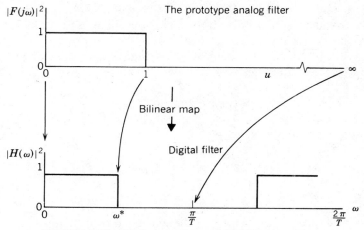

FIGURE 7.27. The frequency map of the bilinear transform.

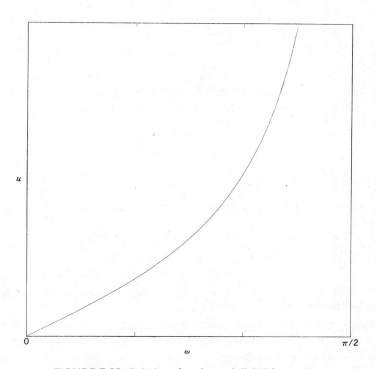

FIGURE 7.28. Relation of analog and digital frequencies.

This prewarping or scaling of the prototype frequency scale is done by replacing s with Ks. Because the bilinear transform is also a change of variables, both can be performed in one step.

If the critical frequency for the prototype filter is u_0 and the desired critical frequency for the digital filter is ω_0, the two frequency responses are related by

$$F(ju_0) = H(\omega_0) = F^*. \tag{7.111}$$

The prewarping scaling is given by

$$u_0 = \frac{2}{T} \tan\left(\frac{\omega_0 T}{2}\right). \tag{7.112}$$

Combining the prewarping scale and the bilinear transformation gives

$$u_0 = \frac{2K}{T} \tan\left(\frac{\omega_0 T}{2}\right).$$

Solving for K and combining with (7.106) give

$$s = \frac{u_0}{\tan(\omega_0 T/2)} \frac{z-1}{z+1}. \tag{7.113}$$

All of the optimal filters developed in Section 7.2 and most other prototype filters are designed with a normalized critical frequency $u_0 = 1$. Recall that ω_0 is in radians per second. Most specifications are given in terms of frequency f in Hertz, which is related to ω or u by

$$\omega = 2\pi f.$$

Care must be exercised with the elliptic function filter when there are two critical frequencies that determine the transition region. Both frequencies must be prewarped.

The characteristic of the bilinear transform are the following:

1. The order of the digital filter is the same as the prototype filter.
2. The left-half of the s plane is mapped into the unit circle on the z plane, which means stability is preserved.
3. Optimal approximations to piecewise constant prototype filters, such as the four cases in Section 7.2, transform into optimal digital filters.
4. The cascade of sections designed by the bilinear transform is the same as that obtained by transforming the total system.

The bilinear transform is probably the most frequently used method for converting a prototype Laplace transform transfer function into a digital transfer function. It is the one used in most popular filter design programs,[10-12] because optimality is preserved. The maximally flat prototype is transformed into a maximally flat digital filter. This property only holds for approximations to piecewise constant, ideal frequency responses, because the frequency warping does not change the shape of a constant. If the prototype is an optimal approximation to a differentiator or to a linear-phase characteristic, the bilinear transform will destroy the optimality. Those approximations have to be made directly in the digital frequency domain.

Example 7.5. Design with The Bilinear Transformation
    To illustrate the bilinear transformation, we convert the third-order Butterworth low-pass filter designed in Example 7.1 into a digital filter. The prototype filter transfer function is

$$F(s) = \frac{1}{(s + 1)(s^2 + s + 1)}. \tag{7.114}$$

The prototype analog filter has a pass-band edge at $u_0 = 1$. We assume a data rate of 1000 samples/s, corresponding to $T = 0.001$ seconds. If the desired digital pass-band edge is $f_0 = 200$ Hz, then $\omega_0 = (2\pi)(200)$ rad/s, and the total prewarped bilinear transformation from (7.113) is

$$s = 1.376382 \, \frac{z - 1}{z + 1}.$$

The digital transfer functionin (7.114) becomes

$$H(z) = \frac{0.09853116(z + 1)^3}{(z - 0.158384)(z^2 - 0.418856z + 0.355447)}. \tag{7.115}$$

Note the locations of the poles and zeros in the z plane. Zeros at infinity in the s plane always map into the $z = -1$ point. The examples in Figs. 7.1 and 7.2 illustrate a third-order elliptic function filter designed with the bilinear transform.

7.3.3 Frequency Transformations

For the design of high-pass, bandpass, and band-rejection filters, a particularly powerful combination consists of using the frequency transformations described in Section 7.2.10 together with the bilinear transformation. When using this combination, be careful to scale the specifications properly. This scaling procedure is illustrated by considering the steps in the design of a bandpass filter:

1. First, the lower and upper digital band-edge frequencies are specified as ω_1 and ω_2 or ω_1, ω_2, ω_3, and ω_4 if an elliptic function approximation is used.
2. These frequencies are prewarped. Equation (7.112) is used to give the band edges of the prototype bandpass analog filter.
3. These frequencies are converted into a single band edge ω_p or ω_s for the Butterworth and Chebyshev approximations and into ω_p and ω_s for the elliptic function approximation of the prototype low-pass filter by using (7.97) and (7.98).
4. The low-pass filter is designed for this ω_p and/or ω_s by using one of the four approximations in Sections 7.2.1 through 7.2.8 or some other method.
5. This low-pass analog filter is converted into a bandpass analog filter with the frequency transformation (7.99).
6. The bandpass analog filter is then transformed into the desired bandpass digital filter with the bilinear transformation (7.106).

This procedure is used in the design Program 9 in the appendix.

In the design of a bandpass elliptic function filter, four frequencies must be specified: the lower stop-band edge, the lower pass-band edge, the upper pass-band edge, and the upper stop-band edge. All four must be prewarped to the equivalent analog values. A problem occurs when the two transition bands of the bandpass filter are converted into the single transition band of the low-pass prototype filter. In general, they will be inconsistent; therefore, the narrower of the two transition bands should be used to specify the low-pass filter. The same problem occurs in designing a band-rejection elliptic function filter. Program 9 should be studied to understand how this is carried out.

An alternative to the process of converting a low-pass analog filter into a bandpass analog filter into a digital filter is to first convert the prototype low-pass analog filter into a low-pass digital filter and then to make the conversion into a bandpass filter. If the prototype digital filter transfer function is $H_p(z)$ and the frequency transformation is $f(z)$, the desired transformed digital filter is described by

$$H(z) = H_p(f(z)).$$

Since the frequency responses of both $H(z)$ and $H_p(z)$ are obtained by evaluating them on the unit circle in the z plane, $f(z)$ should map the unit circle onto the unit circle ($|z| = 1 \Rightarrow |f(z)| = 1$). Both $H(z)$ and $H_p(z)$ should be stable; therefore, $f(z)$ should map the interior of the unit circle into the interior of the unit circle ($|z| < 1 \Rightarrow |f(z)| < 1$). If $f(z)$ were viewed as a filter, it would be an all-pass filter with a unity magnitude frequency response of the form

$$f(z) = \frac{p(z)}{z^n p(1/z)} = \frac{a_n z^n + a_{n-1} z^{n-1} + \cdots + a_0}{a_0 z^n + a_1 z^{n-1} + \cdots + a_n}.$$

The prototype digital low-pass filter is usually designed with band edges at $\pm \pi/2$. Determining the frequency transformation then becomes the problem of solving the $n+1$ equations

$$f(e^{j\omega_i}) = e^{\pm j\pi/2} = (-1)^i j$$

for the unknown a_k where $i = 0, 1, 2, \ldots, n$ and the ω_i are the band edges of the desired transformed frequency response put in ascending order. The resulting simultaneous equations have a special structure that allow a recursive solution. Details of this approach can be found in reference 4.

This approach is extremely general and allows multiple pass bands of arbitrary width. If elliptic function approximations are used, only one of the transition bandwidths can be independently specified. If more than one pass band or band rejection is desired, $f(z)$ will be higher than second order, and therefore the transformed transfer function $H(f(z))$ will have to be factored by a root finder.

To illustrate the results of using transform methods to design filters, we give three examples, which are designed with Program 9.

Example 7.6. Design of a Chebyshev High-Pass Filter
The specifications are given for a high-pass Chebyshev frequency response with a pass-band edge at $f_p = 0.3$ Hz with a sampling rate of 1 sample/s. The order is set at $N = 5$ and the pass-band ripple at 0.91515 dB. The transfer function

$$H(\omega) = \frac{(z-1)(z^2 - 2z + 1)(z^2 - 2z + 1)}{(z + 0.64334)(z^2 + 0.97495z + 0.55567)(z^2 + 0.57327z + 0.83827)}.$$

The frequency-response plot is given in Fig. 7.29.

Example 7.7. Design of an Elliptic Function Bandpass Filter
This filter requires a bandpass frequency response with an elliptic function approximation. The maximum pass-band ripple is 1 dB, the minimum stop-band attenuation is 30 dB, the lower stop-band edge $f_1 = 0.19$ Hz, the lower pass-band edge $f_2 = 0.2$ Hz, the upper pass-band edge $f_3 = 0.3$ Hz, and the upper stop-band edge $f_4 = 0.31$ Hz, with a sampling rate of 1 sample/s. The design program calculated a required prototype order of $N = 5$ and, therefore, a total order of 10. Figure 7.30 shows the frequency-response plot.

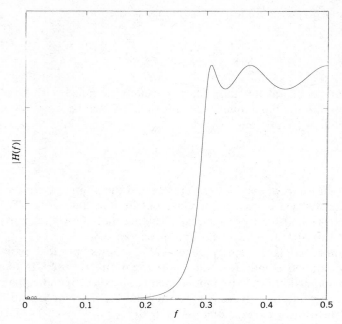

FIGURE 7.29. Fifth-order Chebyshev highpass filter.

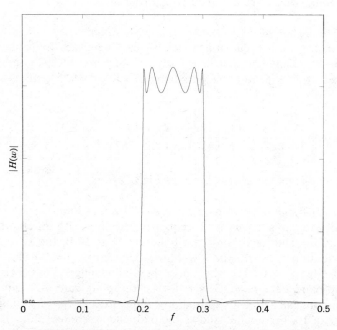

FIGURE 7.30. Tenth-order elliptic function bandpass filter.

216

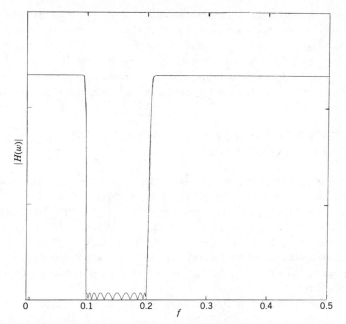

FIGURE 7.31. Twenty second-order inverse Chebyshev band-rejection filter.

Example 7.8. Design of an Inverse Chebyshev Band-Rejection Filter
    The specifications are given for a band-rejection inverse Chebyshev frequency response with band edges at $f_s = 0.1$ and $0.2\,\text{Hz}$ with a sampling rate of 1 sample/s. The prototype order is set at $N = 11$, and the minimum stop-band attenuation at $30\,\text{dB}$. The frequency-response plot is given in Fig. 7.31.

Summary

This section described the two most popular and useful methods for transforming a prototype analog filter into a digital filter. The analog frequency variable is used because literature on analog filter design exists, but, more imporantly, many approximation theories are more straightforward in terms of the Laplace transform variable than the z transform variable. The impulse-invariant method is particularly valuable when time-domain characteristics are important. The bilinear transform method is the most common when frequency-domain performance is the main interest. The bilinear transformation warps the frequency scale; therefore, the digital band edges must be prewarped to obtain the necessary band edges for the analog filter design. Formulas that transform the analog prototype filters into the desired digital filters and for prewarping specifications were derived.
    The use of frequency transformations to convert low-pass filters into high-

pass, bandpass, and band-rejection filters was discussed as a particularly useful combination with the bilinear transformation. These transformations are implemented in Program 9 and design examples from this program were shown.

In some cases no analytic results are possible, or the desired frequency response is not piecewise constant. Transformation methods are then not appropriate. Direct methods for these cases are developed in the next section.

7.4. DIRECT FREQUENCY-DOMAIN IIR FILTER DESIGN METHODS

The preceding design methods have been based on designing an analog prototype filter and then converting it to a digital filter. This approach is appropriate for the class of approximations where analytic solutions are possible; it is not appropriate for many others. The rest of this chapter develops methods that directly design the desired digital filter. Most approaches are extensions of methods used for FIR filters, but they are more complicated for the IIR case where rational approximation rather than polynomial approximation is being performed.

This section develops a frequency-sampling design method such that the frequency response of the IIR filter will pass through the given samples of a desired response. Since an IIR filter cannot have linear phase, the sampled response must contain both magnitude and phase. The extension of the frequency-sampling method to an LS error approximation is not as simple as for the FIR filter. The method given here uses a criterion based on the equation error rather than on the more common error between the actual and desired frequency responses. Nevertheless, it is a useful noniterative design method. Finally, a general discussion of iterative design methods for LS frequency-response error is given.

7.4.1 Frequency-Sampling Design of IIR Filters

The method for calculating samples of the frequency response of an IIR filter presented in Section 6.2 can be reversed to design a filter in much the same way as it was for the FIR filter in Section 3.1. The z transform transfer function for an IIR filter is

$$H(z) = \frac{B(z)}{A(z)} = \frac{b_0 + b_1 z^{-1} + \cdots + b_M z^{-M}}{1 + a_1 z^{-1} + \cdots + a_N z^{-N}}. \tag{7.116}$$

The frequency response of the filter is given by setting $z = e^{-j\omega}$, as shown in Section 6.2. Using the notation

$$H(\omega) = H(z)|z = e^{-j\omega}, \tag{7.117}$$

we choose equally spaced samples of the frequency response so that the number of samples is equal to the number of unknown coefficients in (7.116). These $L + 1 = M + N + 1$ samples of this frequency response are given by

$$H_k = H(\omega_k) = H\left(\frac{2\pi k}{L + 1}\right), \qquad k = 0, 1, \ldots, L, \qquad (7.118)$$

and can be calculated from the length-$(L + 1)$ DFTs of the numerator and denominator as given in (6.6).

$$H_k = \frac{\mathrm{DFT}\{b_n\}}{\mathrm{DFT}\{a_n\}} = \frac{B_k}{A_k}, \qquad (7.119)$$

where the indicated division is term by term for each value of k. Multiplication of both sides of (7.119) by A_k gives

$$B_k = H_k A_k. \qquad (7.120)$$

If the length-$(L + 1)$ inverse DFT of H_k is denoted by the length-$(L + 1)$ sequence h_n, equation (7.120) becomes cyclic convolution, which can be expressed in matrix form by

$$
\begin{bmatrix} b_0 \\ b_1 \\ b_2 \\ \vdots \\ b_M \\ 0 \\ \vdots \\ 0 \end{bmatrix}
=
\begin{bmatrix}
h_0 & h_L & h_{L-1} & \cdots & h_2 & h_1 \\
h_1 & h_0 & h_L & & & h_2 \\
h_2 & h_1 & h_0 & & & \\
\vdots & & & & & \vdots \\
& & & & & \\
h_L & & & \cdots & & h_0
\end{bmatrix}
\begin{bmatrix} 1 \\ a_1 \\ a_2 \\ \vdots \\ a_N \\ 0 \\ \vdots \\ 0 \end{bmatrix}. \qquad (7.121)
$$

Note that the h_n in (7.121) are not the impulse-response values of the filter as in (6.2). A more compact matrix notation of (7.121) is

$$
\begin{bmatrix} \mathbf{b} \\ \mathbf{0} \end{bmatrix} = [H]\begin{bmatrix} \mathbf{a} \\ \mathbf{0} \end{bmatrix}, \qquad (7.122)
$$

where H is $(L + 1)$ by $(L + 1)$, $\mathbf{b}$ is length $(M + 1)$, and $\mathbf{a}$ is length $(N + 1)$. Because the lower $L - N$ terms of the right-hand vector of (7.121) are zero, the H matrix can be reduced by deleting the rightmost $L - N$ columns to give H_0, which transforms (7.122) to

$$
\begin{bmatrix} \mathbf{b} \\ \mathbf{0} \end{bmatrix} = [H_0][\mathbf{a}]. \qquad (7.123)
$$

Because the first element of **a** is 1, it is partitioned to remove the unity term, and the remaining length-n vector is denoted **a***. The simultaneous equations represented by (7.123) are uncoupled by further partitioning of the H matrix, as shown in

$$\begin{bmatrix} \mathbf{b} \\ \hline \mathbf{0} \end{bmatrix} = \begin{bmatrix} & H_1 & \\ \hline \mathbf{h}_1 & \vdots & H_2 \end{bmatrix} \begin{bmatrix} 1 \\ \hline \mathbf{a}^* \end{bmatrix}, \tag{7.124}$$

where H_1 is $(M + 1)$ by $(N + 1)$, $\mathbf{h}_1$ is length $(L - M)$, and H_2 is $(L - M)$ by N. The lower $L - M$ equations are written

$$\mathbf{0} = \mathbf{h}_1 + H_2 \mathbf{a}^*$$

or

$$\mathbf{h}_1 = -H_2 \mathbf{a}^*, \tag{7.125}$$

which must be solved for **a***. The upper $M + 1$ equations of (7.124) are written

$$\mathbf{b} = H_1 \mathbf{a}, \tag{7.126}$$

which allows **b** to be calculated.

If $L = N + M$, H_2 is square. If H_2 is nonsingular, (7.125) can be solved exactly for the denominator coefficients in **a***, which are augmented by the unity term to give **a**. From (7.126) we find the numerator coefficients in **b**.

Note that any order numerator and denominator can be prescribed. If the filter is an FIR filter, **a** is unity and **a*** does not exist. Under these conditions (7.126) states that $b_n = h_n$, which is one of the cases of FIR frequency sampling covered in Section 3.1. Also note that there is no control over the stability of the designed filter for this method.

Summary

This section developed and analyzed an interpolation design method. The frequency-domain specifications were converted to the time domain by the DFT. A matrix partitioning allowed the solution for the numerator coefficients to be uncoupled from the solution of the denominator coefficients. The DFT prevents the possibility of unequally spaced frequency samples, which was possible for FIR filter design.

The frequency-sampling design of IIR filters is somewhat more complicated than for FIR filters because of the requirement that H_2 be nonsingular. As for the FIR filter, the samples of the desired frequency response must satisfy the conditions to ensure that h_n are real. The power of this method is its ability to interpolate arbitrary magnitude and phase specifications. In contrast to most

direct IIR design methods, this method does not require any iterative optimization with the accompanying convergence problems.

As with the FIR version, this design approach is an interpolation method
rather than an approximation method, so it sometimes gives poor performance
between the interpolation points. This usually happens when the desired
frequency-response samples are not consistent with what an IIR filter can
achieve. One solution to this problem is the same as for the FIR case in Section
3.2: the use of more frequency samples than the number of filter coefficients, and
the definition of an approximation error function that can be minimized. No
restriction will guarantee stable filters. If the frequency-response samples are
consistent with an unstable filter, that is what will be designed.

7.4.2 Discrete Least Squared Equation-Error
IIR Filter Design

To obtain better practical filter designs, we extend the interpolation scheme of
the previous section to give an approximation design method.[26,27] Note that the
method developed here minimizes an equation-error measure and not the usual
frequency-response error measure.

The number of frequency samples specified, $L + 1$, will be made larger than
the number of filter coefficients, $M + N + 1$. This means that H_2 is rectangular,
and therefore (7.125) cannot, in general, be satisfied. To formulate an approximation problem, we introduce a length-$(L + 1)$ error vector e in (7.121) and
(7.123) to give

$$\begin{bmatrix} b \\ 0 \end{bmatrix} = [H_0][a] + [e]. \tag{7.127}$$

Equation (7.125) becomes

$$h_1 - e = -H_2 a^*, \tag{7.128}$$

where now H_2 is rectangular with $L - M > N$. Using the same methods as in
Section 3.2 to derive (3.19), we minimize the error e in a LS error sense by solving
the normal equations

$$H_2^T h_1 = -H_2^T H_2 a^*. \tag{7.129}$$

If the equations are not singular, the solution is

$$a^* = -[H_2^T H_2]^{-1} H_2^T h_1. \tag{7.130}$$

The numerator coefficients are found by the same techniques as in (7.126):

$$b = H_1 a,$$

which makes the upper $M + 1$ terms in **e** zero and the total squared error a minimum.

As noted in Section 3.2 on LS error design of FIR filters, (7.129) is often numerically ill conditioned, and (7.130) should not be used to solve for **a***. Special algorithms, such as those contained in LINPACK,[9] should be employed; they were used in the programs in the appendix.

The error **e** defined in (7.127) can better be understood by considering the frequency-domain formulation. Taking the DFT of (7.127) gives

$$B_k = H_k A_k + E_k. \tag{7.131}$$

E is the error produced by trying to satisfy (7.120) when the equations are overspecified. Equation (7.131) can be reformulated in terms of $\mathscr{E}$, the difference between the frequency response samples of the designed filter and the desired response samples, by dividing (7.131) by A_k:

$$\mathscr{E}_k = \frac{B_k}{A_k} - H_k = \frac{E_k}{A_k}. \tag{7.132}$$

$\mathscr{E}$ is the error in the solution of the approximation problem, and E is the error in the equations defining the problem. The usual statement of a frequency-domain approximation problem is in terms of minimizing some measure of $\mathscr{E}$, but that results in solving nonlinear equations. The design procedure developed in this section minimizes the squared error E; thus only linear equations need to be solved. There is an important relation between these problems. Equation (7.132) shows that minimizing E is the same as minimizing $\mathscr{E}$ weighted by A. However, A is unknown until after the problem is solved.

Although this method is posed as a frequency-domain design method, the methods of solution for both the interpolation problem and the LS equation-error problem are similar to the time-domain Prony method,[39] discussed in Section 7.5.

Numerous modifications can be made to this method. If the desired frequency response is close to what can be achieved by an IIR filter, this method will give a design approximately the same as that of a true LS solution-error method. It can be shown that $E = 0 \Leftrightarrow \mathscr{E} = 0$. In some cases improved results can be obtained by estimating A_k and using that as a weight on E to approximate minimizing $\mathscr{E}$. There are iterative methods based on solving (7.130) and (7.126) to obtain values for A_k. These values are used as weights on E to solve for a new set of A_k used as a new set of weights to solve again for A_k.[26,27] The solution of (7.130) and (7.126) is sometimes used to obtain starting values for iterative optimization algorithms that need good starting values for convergence.

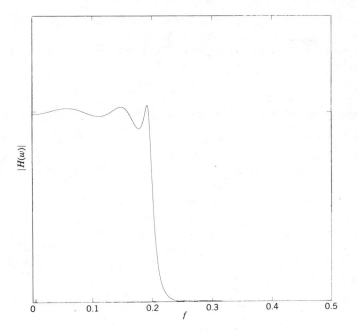

FIGURE 7.32. Sixth-order least-squared equation-error IIR filter.

Example 7.9. Design of Least Square Equation-Error IIR Filter
    To illustrate this design method, we designed a sixth-order low-pass filter with 41 frequency samples to approximate. The magnitude of those less than 0.2 Hz is one and of those greater than 0.2 is zero. The phase was experimentally adjusted to result in a good magnitude response. The design was performed with Program 10 and the frequency response is shown in Fig. 7.32.

Summary

This section gave an LS error approximation method to design IIR filters. By using an equation-error rather than a solution-error criterion, we obtained a problem requiring only the solution of simultaneous linear equations.
    Like the FIR filter version, the IIR frequency sampling design method and the LS equation-error extension can be used for complex approximation and, therefore, can design with both magnitude and phase specifications.
    If the desired frequency-response samples are close to what an IIR filter of the specified order can achieve, this method will produce a filter very close to what a true LS error method would. However, when the specifications are not consistent with what can be achieved and the approximating is large, the results

can be very poor or, in some cases, unstable. It is particularly difficult to set realistic phase-response specifications. With this method it is even more important to have a design environment that will allow an easy trial-and-error procedure.

7.4.3 Least Squared Error Frequency-Domain Design

Practical problems occur in the design of a filter to separate signals according to their energy. Because the energy content of a signal is the integral or sum of the square of the signal, a mean squared error measure is natural. Unfortunately, for the IIR filter design problem, the optimization procedure is nonlinear. This fact was pointed out in the last section, where the equation error was used in order to have a linear problem.

Because of the nonlinear nature of the LS error minimization, the method of solution becomes dependent on the desired frequency response, and therefore there is no single method for design. The mean squared error for magnitude approximation is defined as

$$q(x) = \sum_{i=0}^{L} (|H(\omega_i)| - |H_d(\omega_i)|)^2, \tag{7.133}$$

where x is a vector of filter parameters chosen to minimize q, and the error is sampled at $L + 1$ frequencies ω_i. Steiglitz[28] chose the parameter vector x to be the coefficients of a cascade structure in order to best fit an iterative optimization scheme. He applied a standard optimization algorithm—the Fletcher-Powell method—to the minimization of (7.133). Other methods more directly related to a squared error measure can also be used.[8,29]

Practical difficulties exist in solving this approximation problem. In some cases local minima rather than the global minimum are found. In other cases convergence of the minimization algorithm is slow or does not occur at all. Numerical problems can result from ill-conditioned equations, and there is no guarantee that the designed filter will be stable.

Choosing a desired frequency-response function $H_d(\omega)$ so that the optimum approximation does not have a large error is important. It often means not having an abrupt discontinuity between the pass band and stop band. The techniques discussed in Section 3.2.2.1 are also applicable here.

Another factor is starting the iterative optimization algorithm with a set of coefficients in x that is close to the optimum. That can be done by using the frequency-sampling method or the method of Section 7.4.2 to give a design that can be used to start an LS algorithm. Because the error defined in (7.133) is in terms of magnitudes, an unstable design can be converted to a stable one by moving the unstable pole at a radius r in the z plane to a radius $1/r$. This change does not affect the magnitude frequency response, but it does stabilize the effect of that pole.[28]

A generalization of the idea of a squared error measure is defined by raising the error to the p power, where p is a positive integer. This error is defined by

$$q(x) = \sum_{i=0}^{L} (|H(\omega_i) - H_d(\omega_i)|)^p. \tag{7.134}$$

Deczky[36] developed this approach and used the Fletcher–Powell method to minimize (7.134). He also applied this method to the approximation of a desired group delay function. An important characteristic of this formulation is that the solution approaches the Chebyshev or min-max solution as p becomes large. A program for this design method is given in reference 10.

7.4.4 The Chebyshev Error Criterion for IIR Filter Design

The error measure that often best meets filter design specifications is the maximum error in the frequency response that occurs over a band. The filter design problem becomes the problem of minimizing the maximum error (the min-max problem).

One approach to this error minimization, by Deczky, minimizes the p power error of (7.134) for large p. Generally, $p = 10$ or greater approximates a Chebyshev result.[10,30] Dolan and Kaiser[10] use a penalty function approach.

Linear programming can be applied to this error measure[31–33] by linearizing the equations in much the same way as in (7.129).[1] In contrast to the FIR case this can be a practical design method because the order of a practical IIR filter is generally much lower than for an FIR filter. A scheme called *differential correction* has also proven to be effective[34,35].

Although the rational approximation problem is nonlinear, an application of the Remes exchange algorithm can be implemented[36–38]. Since the zeros of the numerator of the transfer function mainly control the stop-band characteristics of a filter, and the zeros of the denominator mainly control the pass band, the effects of the two are somewhat uncoupled. An application of the Remes exchange algorithm, alternating between the numerator and denominator, gives an effective method for designing IIR filters with a Chebyshev error criterion.[37] If the orders of the numerator and denominator are the same and the desired filter is an ideal low-pass filter, the Remes exchange should give the same result as the elliptic function filter in Section 7.2.4. However, this approach allows a numerator or denominator of any order to be set and pass band of any shape to be approximated. In some cases a filter whose denominator has lower order than its numerator produces fewer required multiplications than an elliptic-function filter.[37,38]

7.5 PRONY'S METHOD FOR TIME-DOMAIN DESIGN OF IIR FILTERS

This section addresses the problem of designing an IIR digital filter with a prescribed time-domain response. Most formulations of time-domain design of IIR filters give nonlinear equations for the same reasons as for frequency-domain design. Prony, in 1790, derived a special formulation to analyze elastic properties of gases, which produced linear equations. A more general form of Prony's method can be applied to the IIR filter design by using a matrix description.[39]

The transfer function of an IIR filter is given by

$$H(z) = \frac{B(z)}{A(z)} = \frac{b_0 + b_1 z^{-1} + \cdots + b_M z^{-M}}{1 + a_1 z^{-1} + \cdots + a_N z^{-N}}, \tag{7.135}$$

and the impulse response $h(n)$ is related to $H(z)$ by the z transform.

$$H(z) = \sum_{n=0}^{\infty} h(n) z^{-n}.$$

Equation (7.135) can be written as

$$B(z) = H(z)A(z), \tag{7.136}$$

which is the z transform version of convolution. This convolution can be written as a matrix multiplication. Using the first $K + 1$ terms of the impulse response, we write

$$
\begin{bmatrix} b_0 \\ b_1 \\ b_2 \\ \vdots \\ b_M \\ 0 \\ \vdots \\ 0 \end{bmatrix}
=
\begin{bmatrix} h_0 & 0 & 0 & \cdots & 0 \\ h_1 & h_0 & 0 & & \\ h_2 & h_1 & h_0 & & \\ \vdots & & & & \vdots \\ h_M & & & & \\ \vdots & & & & \\ h_K & & \cdots & & h_{K-N} \end{bmatrix}
\begin{bmatrix} 1 \\ a_1 \\ a_2 \\ \vdots \\ a_N \end{bmatrix}. \tag{7.137}
$$

To uncouple the calculations of the a_n and the b_n, we partition the matrices in the same way as in (7.124) to give

$$
\begin{bmatrix} \mathbf{b} \\ \hline \mathbf{0} \end{bmatrix}
=
\begin{bmatrix} H_1 \\ \hline \mathbf{h}_1 & H_2 \end{bmatrix}
\begin{bmatrix} 1 \\ \hline \mathbf{a}^* \end{bmatrix}, \tag{7.138}
$$

where **b** is the vector of the $M + 1$ numerator coefficients of (7.135), **a*** is the vector of the N denominator coefficients ($a_0 = 1$), $\mathbf{h}_1$ is the vector of the last $K - M$ terms of the impulse response, H_1 is the $(M + 1)$-by-$(N + 1)$ partition of (7.137), and H_2 is the $(K - M)$-by-N remaining part. The lower $K - M$ equations are written

$$0 = \mathbf{h}_1 + H_2\mathbf{a}^*$$

or

$$\mathbf{h}_1 = -H_2\mathbf{a}^*, \tag{7.139}$$

which must be solved for **a***, the denominator coefficients in (7.135). The upper $M + 1$ equations of (7.138) are written

$$\mathbf{b} = H_1\mathbf{a}, \tag{7.140}$$

which allow **b**, the numerator coefficients of the transfer function (7.135), to be calculated.

If $K = M + N$, H_2 is square. If H_2 is not singular, (7.139) can be solved for **a**, and **b** can be calculated from (7.140). For this case there are $M + N + 1$ unknown coefficients, and therefore the same number of impulse-response terms can be matched. If H_2 is singular, (7.139) may have many solutions, in which case $h(n)$ can be generated by a lower-order system.

Although Prony's method, applied to the time-domain design problem here, is similar to the solution of the frequency-sampling design problem, there are important differences. In (7.120) the IDFT is used to obtain the matrix of (7.121), which is cyclic convolution. Equation (7.137) is noncyclic convolution, and the $K + 1$ terms of $h(n)$, used to form H, result from a truncation of the infinitely long sequence.

Because the basic Prony method is an interpolation scheme to design a filter that exactly produces the first $K + 1$ terms of the specified $h(n)$, it says nothing about $h(n)$ for $n > K$. To control $h(n)$ over a larger range of n, we pose an approximation problem. We define an equation-error vector for (7.138)

$$\begin{bmatrix} \mathbf{b} \\ \hline \mathbf{0} \end{bmatrix} + \begin{bmatrix} \mathbf{e} \end{bmatrix} = \begin{bmatrix} H_1 \\ \hline \mathbf{h}_1 \; \vdots \; H_2 \end{bmatrix}\begin{bmatrix} 1 \\ \hline \mathbf{a}^* \end{bmatrix}. \tag{7.141}$$

If $K > M + N$, we cannot solve exactly (7.141), but we can find **b** and **a** that will minimize the norm of **e** by using the same methods as for (3.19). The normal equations of the lower part of (7.141) are

$$H_2^T\mathbf{h}_1 = -H_2^T H_2\mathbf{a}^*. \tag{7.142}$$

If H_2 has full rank, (7.142) can be solved for $\mathbf{a}^*$. This solution minimizes the lower part of $\mathbf{e}$, and $\mathbf{b} = H_1\mathbf{a}$ gives zero error for the upper part.

If the solution error is defined as the difference between the desired impulse response and the actual impulse response by

$$e(n) = h(n) - h_d(n),$$

the length-$(K + 1)$ solution-error vector e is related to the equation-error vector $\mathbf{e}$ of (7.141) by

$$\mathbf{e} = A e, \tag{7.143}$$

where A is a $(K + 1)$-by-$(K + 1)$ convolution matrix formed from the $a(n)$ coefficients. Prony's method minimizes $\|\mathbf{e}\|$, which is a weighted version of $\|e\|$.

Various modifications can be made to the form presented of Prony's method. After the denominator is found by minimizing the equation error, the numerator can be found by minimizing the solution error. It is possible to mix the exact and approximate methods of (7.138) and (7.141). The details can be found in references 39–41.

Several modifications to Prony's method have been made to use it to minimize the solution error. Most of these iteratively minimize a weighted equation error with Prony's method and update the weights from the previous determination of $\mathbf{a}$.[42,43]

If an LS error, time-domain approximation is the desired result, a minimization technique can be applied directly to the solution error. The most successful method seems to be the Gauss-Newton algorithm with a step-size control. Combined with Prony's method to find starting parameters, it is an effective design tool.

7.6 IIR FILTER DESIGN PROGRAMS

Several digital filter design programs are available. Most of the examples in this book were designed by the programs in the appendix. A more user-friendly commercial program is available from ASPI.[11] It has provisions for Butterworth, Chebyshev I and II, and elliptic function filters with low-pass, high-pass, bandpass, and band-rejection forms. The program runs on an IBM or Texas Instruments (PC) and can produce machine language code for the TMS32010 digital signal-processing chip. A somewhat similar program for the same IIR filter types is called DISPRO,[12] which runs on the IBM PC.

Certain IIR designs can be carried out by the large and versatile programs from ILS[14] and ISP.[15] These are general programs and run on mainframe computers, but smaller versions are available that will run on a PC.

The FORTRAN program for the four classical approximations is given as program 9. It is written to closely follow the notation and theory developed in this chapter so that each can help in understanding the other. The algorithms used are the most accurate and efficient known to the authors; however, the inputoutput sections are primitive and would have to be further developed for easy use.

Four programs for IIR filter design are in reference 10. Part of the program DOREDI designs and simulates IIR filters, Deczky's program, Dolan and Kaiser's program, and Steiglitz and Ladendorf's program are all available in FORTRAN.[10] A part of the SIG package from Lawrence Livermore Labs[13] designs IIR filters. Indeed, SIG is a very valuable tool for the signal processor.

It is very instructive to design a variety of filters with different specifications in order to develop insight into their various characteristics. It is best accomplished with an interactive program with graphics output.

Summary

The chapter developed the main approaches to IIR filter design. The theory and design equations for the Butterworth, Chebyshev, inverse Chebyshev, and elliptic function filters were given along with variations to the Butterworth and Chebyshev for arbitrary zero locations. The elliptic function filter was developed in more detail than in most books because of the important nature of its optimality. The frequency-sampling and LS equation-error design methods were covered because of their simplicity and their ability to approximate arbitrary, complex, desired frequency responses. The problems of general LS error design and Chebyshev error design using the Remes algorithm were described and references were given. Finally, time-domain design methods based on Prony's methods were given, and general time-domain LS error methods were described and referenced. After the design of an IIR filter, the transfer function must be realized, and that is the topic of the next chapter.

REFERENCES

[1] L. R. Rabiner and B. Gold, *Theory and Application of Digital Signal Processing*, Englewood Cliffs, NJ: Prentice-Hall, 1975.

[2] A. V. Oppenhim and R. W. Schafer, *Digital Signal Processing*, Englewood Cliffs, NJ: Prentice-Hall, 1975.

[3] F. J. Taylor, *Digital Filter Design Handbook*, New York: Dekker, 1983.

[4] C. T. Mullis and R. A. Roberts, *An Introduction to Digital Signal Processing*, Reading, MA: Addison-Wesley, 1987.

[5] L. R. Rabiner and C. M. Rader, eds., *Digital Signal Processing*, selected reprints, New York: IEEE Press, 1972.

[6] *Digital Signal Processing II*, selected reprints, New York: IEEE Press, 1979.

[7] B. Gold and C. M. Rader, *Digital Processing of Signals*, New York: McGraw-Hill, 1969.

[8]   J. E. Dennis, Jr. and R. B. Schnabel, *Numerical Methods for Unconstrained Optimization and Nonlinear Equations*, Englewood Cliffs, NJ: Prentice-Hall, 1983.

[9]   J. J. Dongarra, J. R. Bunch, C. B. Moler, and G. W. Stewart, *LINPACK Users' Guide*, Philadelphia: SIAM, 1979.

[10]   *Programs for Digital Signal Processing*, New York: IEEE Press, 1979.

[11]   *Digital Filter Design Package, DFDP*, Interactive Software for Digital Filter Design, Version 1.02, Atlanta, GA: Atlanta Signal Processors Inc., 1984.

[12]   J. O'Donnell, *DISPRO v1.0 User's Manual*, Digital Filter Design Software, Wayland, MA: Signix Corp., 1983.

[13]   *SIG: A General Purpose Signal Processing, Analysis, and Display Program*, Livermore, CA: Lawrence Livermore Labs, 1985.

[14]   *ILS: Interactive Signal Processing Software*, Goleta, CA: Signal Technology, Inc.

[15]   *I*S*P: The Interactive Signal Processor*, Bedford, MA: Bedford Research.

[16]   C. S. Burrus and T. W. Parks, *DFT/FFT and Convolution Algorithms*, New York: Wiley-Interscience, 1985.

[17]   D. A. Calahan, *Modern Network Synthesis*, Vol. I, *Approximation*, New York: Hayden, 1964.

[18]   M. E. Van Valkenburg, *Analog Filter Design*, New York: Holt, Rinehart & Winston, 1982.

[19]   L. Weinberg, *Network Analysis and Synthesis*, New York: McGraw-Hill, 1962.

[20]   B. D. Rakovich and V. B. Litovski, "Monotonic Passband Low-Pass Filters with Chebyshev stopband Attenuation," *IEEE Trans. ASSP* **ASSP-22** 39–45 (1974).

[21]   M. Abramowitz and I. A. Stegun, eds., *Handbook of Mathematical Functions*, National Bureau of Standards, 1964, Washington, D. C.: Chaps. 16 and 17 (L. M. Milne-Thomson) pp. 571, 574, 579, 598, 599; reprinted by Dover, 1965.

[22]   R. Bulirsch, "Numerical Calculation of Elliptic Integrals and Elliptic Functions," *Numer. Math.* **7**, 78–90 (1965).

[23]   A. H. Gray and J. D. Markel, "A Computer Program for Designing Digital Elliptic Filters," *IEEE Trans. ASSP* **ASSP-24**, 529–538 (1976).

[24]   C. B. Sharpe, "A General Tchebycheff Rational Function," *Proc. IRE* **42**, 454–457 (1954).

[25]   D. Helman, "Tchebycheff Approximations for Amplitude and Delay with Rational Functions," presented at the Symposium on Modern Network Synthesis, Polytechnic Institute of Brooklyn, April 1955; published in *Modern Network Synthesis*, MRI Symposium Series, vol. 5, 1955, pp. 385–402.

[26]   C. K. Sanathanan and J. Koerner, "Transfer Function Synthesis as a Ratio of Two Complex Polynomials," *IEEE Trans. Automatic Control* **AC-8**, 56–58 (1963).

[27]   M. A. Sid-Ahmed, A. Chottera, and G. A. Jullien, "Computational Techniques for Least-Square Design of Recursive Digital Filters," *IEEE Trans. ASSP* **ASSP-26**, 478–480 (1978).

[28]   K. Steiglitz, "Computer-Aided Design of Recursive Digital Filters," *IEEE Trans. Audio Electroacoustics* **18**, 123–129 (1970).

[29]   M. T. Dolan, "Comments on 'On the Approximation Problem for Recursive Digital Filters with Arbitrary Attenuation Curve in the Pass-Band and the Stop-Band'," *IEEE Trans. ASSP* **ASSP-24**, 575–577 (1976).

[30] A. G. Deczky, "Synthesis of Recursive Digital Filters Using the Minimum p-Error Criterion," *IEEE Trans. Audio Electroacoustics* **20**, 257–263 (1972).

[31] P. Thajchayapong and P. J. W. Rayner, "Recursive Digital Filter Design by Linear Programming," *IEEE Trans. Audio Electroacoustics* **21**, 107–112 (1973).

[32] L. R. Rabiner, N. Y. Graham, and H. D. Helms, "Linear programming Design of IIR Digital Filters with Arbitrary Magnitude Function," *IEEE Trans. ASSP* **ASSP-22**, 117–123 (1974).

[33] A. T. Chottera and G. A. Jullien, "A Linear Programming Approach to recursive Digital Filter Design with Linear Phase," *IEEE Trans. Circuits Systems* **CAS-29**, 139–149 (1982).

[34] D. E. Dudgeon, "Recursive Filter Design Using Differential Correction," *IEEE Trans. ASSP* **ASSP-22**, 443–448 (1974).

[35] S. Crosara and G. A. Mian, "A Note on the Design of IIR Filters by the Differential-Correction Algorithm," *IEEE Trans. Circuits Systems* **CAS-30**, 898–903 (1983).

[36] A. G. Deczky, "Equiripple and Minimum (Chebyshev) Approximations for Recursive Digital Filters," *IEEE Trans. ASSP* **ASSP-22**, 98–111 (1974).

[37] H. G. Martinez and T. W. Parks, "Design of Recursive Digital Filters with Optimum Magnitude and Attenuation Poles on the Unit Circle," *IEEE Trans. ASSP* **ASSP-26**, 150–156 (1978).

[38] T. Saramaki, "Design of Optimum Recursive Digital Filters with Zeros on the Unit Circle," *IEEE Trans. ASSP* **ASSP-31**, 450–458 (1983).

[39] C. S. Burrus and T. W. Parks, "Time Domain Design of Recursive Digital Filters," *IEEE Trans. Audio Electroacoustics* **18**, 137–141 (1970).

[40] F. Brophy and A. C. Salazar, "Considerations of the Pade Approximant Technique in the Synthesis of Recursive Digital Filters," *IEEE Trans. Audio Electroacoustics* **21**, 500–505 (1973).

[41] F. Brophy and A. C. Salazar, "Recursive Digital Filter Synthesis in the time Domain," *IEEE Trans. ASSP* **ASSP-22**, 45–55 (1974).

[42] A. G. Evans and R. Fischl, "Optimal Least Squares Time-Domain Synthesis of Recursive Digital Filters," *IEEE Trans. Audio Electroacousticsl* **21**, 61–65 (1973).

[43] K. Steiglitz, "On the Simultaneous Estimation of Poles and Zeros in Speech Analysis," *IEEE Trans. ASSP* **ASSP-25**, 229–234 (1977).

[44] M. S. Bertran, "Approximation of Digital Filters in One and two Dimensions," *IEEE Trans. ASSP* **ASSP-23**, 438–443 (1975).

[45] J A. Cadzow, "Recursive Digital Filter Synthesis Via Gradient Based Algorithms," *IEEE Trans. ASSP* **ASSP-24**, 349–355 (1976).

8

Implementation of Infinite Impulse-Response Filters

All of the analysis of IIR filters in Part III has so far been in terms of linear systems. When the finite word-length effects of overflow and quantization error are considered, the digital filter becomes a *nonlinear* system. It is these nonlinearities of quantization that cause all of the difficulties in the analysis of fixed-point recursive filter implementations.

This chapter begins with a discussion of different ways to implement recursive filters (different structures), with emphasis on second-order blocks. Quantization noise and coefficient quantization errors are analyzed with linear theory. Finally, the instabilities caused by overflow and quantization in a recursive filter are studied.

8.1 RECURSIVE STRUCTURES

When a filter is implemented with a recursive structure, the finite word-length problems become more severe than the problems associated with a nonrecursive filter structure.[1-3] The following two problems are more difficult to analyze for recursive filters than they are for nonrecursive filters.

1. Filter coefficient errors from quantization.
2. Quantization noise and overflow from arithmetic operations.

In addition to the effects of quantization of the coefficients and of finite-precision arithmetic discussed for nonrecursive filters in Chapter 5, two new problems are caused by the feedback in a recursive filter.

1. Small-scale limit cycles, which are oscillations caused by the quantization

nonlinearity in the seemingly stable feedback loop. They usually have low amplitude and can often be tolerated.

2. Large-scale limit cycles, which are oscillations caused by overflow in the feedback loop. Their amplitude covers the complete dynamic range of the filter, so these cycles must be prevented.

These problems are especially difficult to analyze for recursive filters. The approximation problem of designing a rational transfer function with quantized coefficients has not been solved. Although overflow leads to errors with nonrecursive filters, it can lead to large-amplitude, sustained oscillations in recursive filters. The whole area of instabilities introduced by finite word-length effects is still a subject of research.[4] Different structures have different characteristics for these effects. That is the reason for examining different implementations of the same transfer function (different structures).

In a recursive digital filter the output is a linear combination of past inputs *and* past outputs. Past outputs are fed back to produce the present output. The difference equation

$$y(n) = \sum_{m=0}^{M} b_m x(n - m) - \sum_{m=1}^{N} a_m y(n - m) \qquad (8.1)$$

shows how the output $y(n)$ is computed for a recursive filter with a transfer function

$$H(z) = \frac{b_0 + b_1 z^{-1} + \cdots + b_M z^{-M}}{1 + a_1 z^{-1} + \cdots + a_N z^{-N}} = \frac{B(z)}{A(z)}. \qquad (8.2)$$

When a recursive digital filter is implemented directly, as in (8.1), errors introduced by quantization of the coefficients can cause significant variation from the desired frequency response. A filter designed to be stable can become unstable after the coefficients are quantized.

8.1.1 Coefficient Sensitivity

When the coefficients in the difference equation (8.1), which implements a recursive digital filter, are quantized, the resulting coefficient errors can cause major changes in the filter characteristics. We can understand the effect of coefficient errors on both the frequency response and the stability by studying how the locations of the poles of the transfer function $H(z)$ in (8.2) change when there are changes in the coefficients a_k of the denominator of $H(z)$.

The transfer function of a recursive filter is a rational function of z, as shown in (8.2). To obtain $H(z)$ in terms of positive powers of z, we rewrite it as

$$H(z) = \frac{z^{N-M}(b_0 z^M + \cdots + b_M)}{z^N + a_1 z^{N-1} + \cdots + a_N} \qquad (8.3)$$

or

$$H(z) = z^{N-M} \frac{\tilde{B}(z)}{\tilde{A}(z)}. \tag{8.4}$$

The denominator polynomial in (8.4) may be written as

$$\tilde{A}(z) = \sum_{k=0}^{N} a_k z^{N-k} = \sum_{m=0}^{N} (z - z_m), \tag{8.5}$$

where $a_0 = 1.0$.

To see how a change in coefficient a_k affects the pole location z_m, consider the Taylor series expansion[4] of $\tilde{A}(z)$ considered as a function of z and a_k: $\tilde{A}(z, a_k)$.[5]

$$\tilde{A}(z_m + \Delta z_m, a_k + \Delta a_k) = \tilde{A}(z_m, a_k) + \Delta a_k \frac{\partial \tilde{A}(z)}{\partial a_k} + \Delta z_m \frac{\partial \tilde{A}(z)}{\partial z_m} + \cdots. \tag{8.6}$$

Assuming that Δa_k and Δz_m compensate to keep $A(z)$ the same, we get

$$\Delta z_m = -\Delta a_k \frac{\partial \tilde{A}(z)/\partial a_k}{\partial \tilde{A}(z)/\partial z_m}. \tag{8.7}$$

Evaluating the partial derivatives in (8.7) gives

$$\frac{\partial \tilde{A}(z)}{\partial a_k} = \frac{\partial}{\partial a_k} \left| \sum_{i=0}^{N} a_i z^{N-i} \right| = z^{N-k}, \tag{8.8}$$

$$\frac{\partial \tilde{A}(z)}{\partial z_m} = \frac{\partial}{\partial z_m} \left| \sum_{j=1}^{N} (z - z_j) \right|_{j \neq m} = -\sum_{j=1}^{N} (z - z_j). \tag{8.9}$$

Evaluating (8.8) and (8.9) at $z = z_m$ gives

$$\Delta z_m = \Delta a_k \left| \frac{z_m^{N-k}}{\prod_{j=1; j \neq m} (z_m - z_j)} \right|. \tag{8.10}$$

The expression for coefficient sensitivity (8.10) leads to several conclusions about recursive filter implementation:

1. The filter is most sensitive to variations of the last coefficient a_N because $N - k$ is zero.
2. Moving the pole z_m closer to the unit circle ($|z| = 1$) increases the sensitivity of the pole location to the variation of a coefficient because the numerator of (8.10) is larger.

3. Coefficient sensitivity increases when the poles are close together because of small values of $z_m - z_j$ in the denominator of (8.10).

When there are sharp transitions in the frequency response (when N is large), it is difficult to have well-separated poles. Thus, for sensitivity reduction, a cascade of several lower-order sections is recommended instead of a direct realization of a high-order filter. In this way it is possible to have well-separated poles within each section and attain reduced sensitivity to coefficient variations within each lower-order section. Fourth- or even higher-order blocks may make sense for implementations where the multiply/accumulate operation is especially easy to do, but generally the transfer function is broken up into second-order sections. The second-order sections are much easier to analyze than higher-order blocks.

Since the coefficient sensitivity, according to conclusion 3, increases when poles are close together, very narrow-band filters are more sensitive to coefficient errors than wide-band filters because the poles are usually clustered around the pass-band region of the z plane.

8.1.2 Second-Order Structures

The sensitivity analysis in Section 8.1.1 indicates that a less-sensitive structure may be obtained by breaking up the transfer function into lower-order sections and connecting these sections in parallel or in cascade. Although higher-order blocks may be attractive in some applications, the second-order section is a good building block to use in parallel or cascade structures. The principles illustrated by the second-order sections described in this section also apply to higher-order sections.

The most direct form for implementing the difference equation

$$y(n) = -a_1 y(n-1) - a_2 y(n-2) + b_0 x(n) + b_1 x(n-1) + b_2 x(n-2) \qquad (8.11)$$

is shown in Fig. 8.1. The direct structure in the figure can be simplified, combining the four delay blocks into two, as shown in Fig. 8.2.

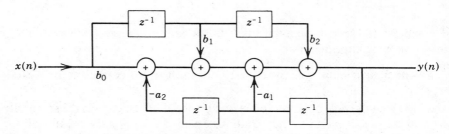

FIGURE 8.1 Direct implementation of a second-order block.

The same difference equation may also be implemented in the transpose structure, shown in Fig. 8.3. It is called the *transpose structure* because it can be obtained from the state-variable or matrix description of Fig. 8.2 by transposing the appropriate matrices,[1] as described in Section 8.1.5. An alternative structure is the coupled form for a second-order block, proposed by Gold and Rader.[6] This structure implements a conjugate pair of poles with real part R and imaginary part $\pm I$, as shown in Fig. 8.4.

Other structures may be used to implement a second-order section. One family of structures may be derived from a state-variable analysis, as described in Section 8.1.5.[1] There are also lattice structures,[7] wave digital filter structures,[8] ladder filters,[9] and many others. Each of these structures may be used to implement low-order blocks and may be combined with other blocks implemented with different structures. The possibilities are endless.

This section has only presented the basic direct, transpose, and coupled structures because they are easy to understand and work quite well when enough bits are available for coefficient and signal representations, so quantizat-

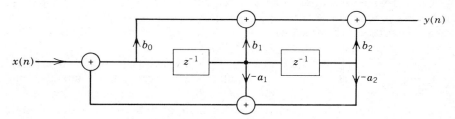

FIGURE 8.2 Direct-form—second-order block.

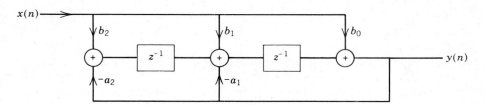

FIGURE 8.3 Transpose Form Second-order Block.

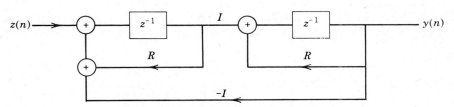

FIGURE 8.4 Coupled-Form for Second Order Block.

ion effects are not serious. When fewer bits are available, more complicated structures less sensitive to quantization errors may be necessary.[1,2]

8.1.3 Cascade Structures

By factoring, we can write the rational transfer function

$$H(z) = \frac{b_0 + b_1 z^{-1} + \cdots + b_M z^{-M}}{1 + a_1 z^{-1} + \cdots + a_N z^{-N}} \tag{8.12}$$

as

$$H(z) = K \prod_{k=1}^{[N/2]} \frac{1 + b_{1k} z^{-1} + b_{2k} z^{-2}}{1 + a_{1k} z^{-1} + a_{2k} z^{-2}}, \tag{8.13}$$

where $[N/2]$ is the smallest integer $\geq N/2$.

Each of the second-order factors in (8.13) can be implemented with one of the structures described in Section 8.1.2, giving a realization of $H(z)$ as a cascade of second-order sections, as shown in Fig. 8.5. If the filter has an odd order, then a first- or third-order section is necessary.

There are many different cascade structures corresponding to different orderings of the $H_k(z)$ blocks and different pairings of the numerator and denominator factors of (8.13). This freedom of ordering and pairing may be used to reduce quantization noise. In the design example in Section 8.5, the zeros are paired with nearby poles to reduce the possibility of a very peaked frequency response for that section. As described by Jackson,[2] second-order sections are ordered so that the section with the poles closest to the unit circle is last. To determine the best pairing and ordering for a particular filter, one must evaluate the quantization noise for all possibilities, using the methods described in Section 8.2.3.

It is possible to use different structures for different sections. For example, those sections with poles near the unit circle can be implemented with structures that have lower sensitivity but may require more computation. Sections with well-separated poles away from the unit circle can be implemented with simpler structures.

A possible advantage of the cascade structure is that unit circle zeros of the overall transfer function can easily be implemented. When the numerator coefficient b_{2k} in (8.13) is equal to unity, the zero for the kth section is on the unit circle. In the cascade structure if one section has a zero on the unit circle, then

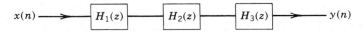

FIGURE 8.5 Cascade Structure.

(except for possible pole-zero cancellation) the entire filter will have a zero on the unit circle.

8.1.4 Parallel Structures

If the denominator of (8.12) has N_r real roots and N_c pairs of complex-conjugate roots, then a partial fraction expansion of (8.12) gives

$$H(z) = \sum_{k=0}^{M-N} p_k z^{-k} + \sum_{k=1}^{N_r} \frac{A_k}{1 - d_k z^{-1}} + \sum_{k=1}^{N_c} \frac{B_k(1 - c_k z^{-1})}{(1 - r_k z^{-1})(1 - r_k^* z^{-1})}. \quad (8.14)$$

When both the real and complex-conjugate poles are grouped in pairs, (8.14) becomes

$$H(z) = \sum_{k=1}^{[N/2]} H_k(z) + \sum_{k=0}^{M-N} p_k z^{-k} \quad (8.15)$$

with

$$H_k(z) = \frac{c_0 k - c_1 k z^{-1}}{1 - a_1 k z^{-1} - a_2 k z^{-2}}. \quad (8.16)$$

The parallel structure is shown in Fig. 8.6 for $M = N$.

In the parallel structure, unlike the cascade structure, reordering the $H_k(z)$ blocks makes no difference; therefore the problem of choosing the order of second-order blocks is avoided. Further, unlike the cascade structure, scaling can be performed for each block independently of the other blocks. A possible disadvantage of the parallel structure is the difficulty of exactly placing zeros on the frequency axis (unit circle). In the cascade structure it is easy to place a zero of the filter on the unit circle by simply placing the zero of one of the cascaded sections on the unit circle. However, in the parallel structure the zeros depend on cancellation of terms in the summation and are more sensitive to coefficient quantization.[2]

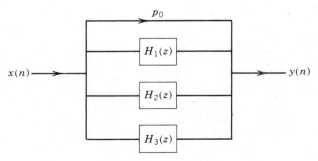

FIGURE 8.6 Parallel Structure.

8.1.5 State-Variable Filter Descriptions

It is often convenient to model a digital filter as a linear time-invariant system with constant coefficient, matrix difference equations called *state* equations.[1]

With the state vector $\mathbf{x}$, single input u, and output y,

$$\mathbf{x}(n + 1) = A\mathbf{x}(n) + Bu(n), \quad y(n) = C\mathbf{x}(n) + Du(n). \tag{8.17}$$

For an Nth-order single-input/single-output system, the sizes of the matrices are

$$
\begin{array}{cc}
A & N \times N \\
B & N \times 1 \\
C & 1 \times N \\
D & 1 \times 1
\end{array}
$$

The transfer function of the system in (8.15) is

$$H(z) = C[zI - A]^{-1}B + D, \tag{8.18}$$

where I is the identity matrix.

Many choices of A, B, C, and D in (8.17) give the same transfer function (8.18). Let

$$A' = M^{-1}AM, \quad B' = M^{-1}B, \quad C' = CM, \quad D' = D. \tag{8.19}$$

The system described in (8.19) has a transfer function

$$H'(z) = C'[zI - A']^{-1}B' + D'. \tag{8.20}$$

Substituting (8.19) into (8.20) gives

$$H'(z) = CM[zI - M^{-1}AM]^{-1}M^{-1}B + D. \tag{8.21}$$

Since

$$[zI - M^{-1}AM]^{-1} = M^{-1}[zI - A]^{-1}M, \tag{8.22}$$

we have $H'(z) = H(z)$.

If the digital filter were a linear system, then all of the infinitely many systems described by (8.20) for different choices of M would all have the same behavior (i.e., would all be equivalent). However, since a digital filter is not a linear system because finite word-length arithmetic is used, different choices of M (different realizations of the filter) will have different properties. The discussion in Section 8.3 describes how to choose M to minimize the effects of quantization noise. A

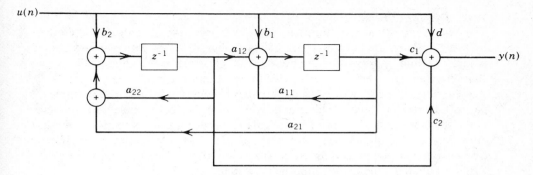

FIGURE 8.7 State-Variable Structure.

block diagram of the state-variable structure is shown in Fig. 8.7 for

$$A = \begin{bmatrix} a_{11} & a_{12} \\ a_{21} & a_{22} \end{bmatrix}, \quad B = \begin{bmatrix} b_1 \\ b_2 \end{bmatrix}, \quad C = c_1 \quad c_2, \quad D = d.$$

The state-variable structure requires much more arithmetic than the simpler direct and transpose structures.

8.1.6 Other Structures

Many other structures for implementing digital filters have been proposed as alternatives to the cascade or parallel connection of second-order blocks. These structures are generally less sensitive to coefficient errors. The ones mentioned here are the lattice[7] and the wave digital filter.[8]

Lattice
The *lattice* structure is widely used for speech synthesis.[7] It is less sensitive to coefficient errors than the direct forms, has a nice interpretation in terms of an acoustic tube, and has a simple way to test for stability. The lattice section in Fig. 8.8 can be connected to other sections to form a higher-order filter.

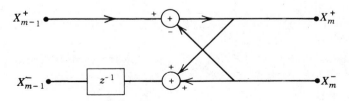

FIGURE 8.8 Lattice Section.

Wave Digital Filters:

The *wave digital filter* structure has been developed from analog *LC* filters by Fettweis.[8] There are several types of wave digital filters with varying computational and storage requirements. The class of wave digital filters is characterized by a very low sensitivity to coefficient errors.

Generally, as the structure of a digital filter becomes more and more complex, part of the load of doing the filtering is lifted from the coefficients and is carried in the structure itself. The more complicated structures, such as the wave digital filter, are capable of operating with very few bits for coefficient representation. Conversely, if 16 or more bits are available for the coefficients and the signal variables, then a simpler structure usually suffices.

Summary

Direct implementation of a high-order recursive filter is not practical with finite word-length fixed-point arithmetic. Low-order blocks can be implemented in the direct form and connected either in cascade or in parallel to construct less-sensitive filter structures.

More complicated filter structures that require more computation are less sensitive to finite word-length effects. Some popular examples of these structures are minimum-noise, state-variable filters, lattice filters, and wave digital filters.

8.2 FINITE WORD-LENGTH EFFECTS

For the minimum computing time or for the most powerful filter that can be computed in a given time, fixed-point arithmetic is usually the best choice. Most signal-processing chips use fixed-point arithmetic for the most efficient use of the limited silicon area available. This section analyzes in detail fixed-point implementations of recursive filters. The finite word-length effects are more complicated and potentially cause more trouble with recursive filters than with nonrecursive filters.

Finite word-length effects are divided into four categories:

1. Filter coefficient errors.
2. Quantization noise and overflow errors in representing signals as fixed-point numbers.
3. Small-scale limit cycles due to the nonlinear quantization characteristics of fixed-point implementations.
4. Large-scale limit cycles due to the nonlinear overflow characteristics of fixed-point implementations.

Each of these aspects of digital filtering requires a different type of analysis.

8.2.1 Coefficient Quantization

As described in Section 8.1.1, a recursive filter is less sensitive to coefficient errors when it is implemented with second-order blocks. Even with a second-order block, there are only a finite number of pole locations because of the coefficient quantization. It may not be possible to place a pole in the exact spot in the z plane specified by a design procedure described in Chapter 7. For example, in digital oscillator design there are limits on the frequencies of oscillation that can be obtained. For very low frequencies (poles near $+1$), especially, there are not many possible pole locations. As shown in Example 8.1, surprisingly few low frequencies are available, even when 16-bit coefficients are used.[9]

Example 8.1 A Digital Oscillator

A digital oscillator has an output $y(n)$ that satisfies the homogeneous difference equation

$$y(n) = 2b_1 y(n-1) - y(n-2). \tag{8.23}$$

The roots of the characteristic equation

$$z^2 - 2b_1 z + 1 \tag{8.24}$$

are located at

$$z_{1,2} = e^{\pm j(2\pi f_0)}, \tag{8.25}$$

where the frequency of oscillation is

$$f_0 = \frac{1}{2\pi} (\cos^{-1} b_1) \tag{8.26}$$

For very low frequencies ($f_0 \cong 0$) the approximation

$$b_1 = \cos(2\pi f_0) \cong 1 - 0.5(2\pi f_0)^2 \tag{8.27}$$

may be used so that the lowest frequencies of oscillation, corresponding to

$$b_1 = 1 - mQ, \qquad m = 1, 2, 3,$$

are approximately

$$f_{0m} = \frac{1}{2\pi} (2mQ)^{1/2}, \qquad m = 1, 2, 3, \tag{8.28}$$

where $Q = 2^{-B+1}$ is the quantization step size. For 16-bit coefficients $B = 16$ and

$$f_{0m} = \frac{\sqrt{m}}{256\pi},$$

$$f_{01} \cong .00124,$$

$$f_{02} \cong .00176,$$

$$f_{03} \cong .00215.$$

The possible frequencies or, equivalently, the possible pole locations are not very dense for regions near $+1$ in the z plane, even with 16-bit coefficients in the difference equation. The reason is that even small changes in the coefficient b_1 cause large changes in the argument of the cosine function in (8.27), because the cosine function is so flat for small values of the angle $2\pi f_0$.

The same quantization of pole locations that is shown in Example 8.1 is also present in recursive filters. Different second-order filter structures have different grids of possible pole and zero locations (see reference 2 for examples).

The problem of optimum design for quantized coefficients is much more difficult for the rational transfer function of the IIR filters than it is in the FIR case, and there are no convenient programs available for designing optimum, quantized coefficient IIR filters. The following trial-and-error approach is useful in practice and should give satisfactory results in most cases.[10]

1. Design the filter and assume no coefficient quantization.
2. Quantize coefficients in the scaled filter and check the frequency response. Also check the pole locations to determine stability of the filter.
3. If the filter in step 2 meets specifications in some bands but not in others, tighten the requirements on the failed bands and relax the weighting on the others. Then repeat step 1. If the filter fails to meet specifications in all bands, increase the order and repeat step 1.

8.2.2 Scaling and Overflow

Scaling is even more important for recursive filters than for nonrecursive filters. For a nonrecursive filter an overflow of the output register only causes an error in the output sample, but for a recursive filter an overflow is fed back and affects many following outputs. For structures that can have large-scale limit cycles, the overflow can set off an oscillation with full-scale amplitude, which completely destroys the usefulness of the filter output for all time after the overflow occurs.

The principles of scaling are the same as for nonrecursive filters. First, the unit-pulse responses (equivalently the frequency responses) are calculated

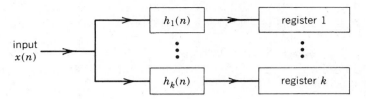

FIGURE 8.9   Scaling Resonses to Registers.

between the input and the various registers (adders) in the filter where overflow might occur (see Fig. 8.9).

The three most useful measures of gain from the input to register k are essentially the same as the measures described in Section 5.3.3. They are repeated here for convenience. We assume that all registers are the same size (e.g., all 16-bit registers) so that the scale factor G_k must be calculated to make the magnitude of the signal at the kth register less than unity to prevent overflow. The scaled unit-pulse response is given as

$$h'_k(n) = \frac{h_k(n)}{G_k}. \tag{8.29}$$

The gain factor is equal to one of the following three measures of the size of $h(n)$.

The l_1 norm of h is of the form

$$\|h\|_1 = \sum_n |h(n)|. \tag{8.30}$$

The Chebyshev norm of the frequency response $H(F)$ is

$$\|H\|_C = \max_F |H(F)|. \tag{8.31}$$

The l_2 norm of h is given as

$$\|h\|_2 = \left[\sum_n h^2(n)\right]^{1/2}. \tag{8.32}$$

If $G_k = \|h\|_1$, then the signal at register k is guaranteed not to overflow. A larger gain occurs (with the resulting smaller quantization noise) if $G_k = \|H\|_C$. This choice of gain only guarantees that the steady-state response of the system to a sine wave will not overflow. Transient signals may occasionally cause overflow. The third choice of the gain factor $G = \|h\|_2$ also allows overflow, but lends itself to a calculation of the probability of overflow[1]. The scaling procedure is described in detail for a second-order section in Example 8.2.

Example 8.2 Scaling a Second-Order Section
   The transpose structure for a second-order filter shown in Fig. 8.10 has a transfer function

$$\frac{Y(z)}{X(z)} = H(z) = \frac{b_0 + b_1 z^{-1} + b_2 z^{-2}}{1 + a_1 z^{-1} + a_2 z^{-2}}. \tag{8.33}$$

This transfer function can be used to calculate the appropriate gain factor to use in scaling to control overflow in the summation at the output. It is also necessary to calculate the transfer functions to the individual internal adders. The output of the first adder is denoted y_1, and the transfer function from the input to y_1 is

$$\frac{Y_1(z)}{X(z)} = H_1(z) = \frac{(b_2 - a_2 b_0) + (b_2 a_1 - a_2 b_1)z^{-1}}{1 + a_1 z^{-1} + a_2 z^{-2}}. \tag{8.34}$$

The output of the second internal adder is denoted y_2, and the transfer function from the input to y_2 is

$$\frac{Y_2(z)}{X(z)} = H_2(z) = \frac{(b_1 - a_1 b_0) + (b_2 - a_2 b_0)z^{-1}}{1 + a_1 z^{-1} + a_2 z^{-2}}. \tag{8.35}$$

These transfer functions (or the corresponding unit-pulse responses) are now used to calculate the gain factors. See the design example in Section 8.4 for a detailed illustration of these scaling principles.

Summary
Scaling is performed by first calculating the transfer function from the input to the register where overflow is possible. Various measures of the effective gain of this transfer function can be used to determine a scale factor to use in reducing the gain, if necessary, so that the possibility of overflow is eliminated, or at least

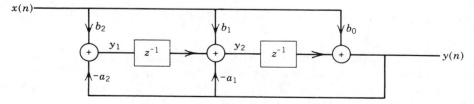

FIGURE 8.10 Transpose Structure Scaling.

limited. The gain should not be reduced any more than necessary in order to preserve the output signal to quantization noise ratio.

8.2.3 Quantization Noise

Multiplying a B_1-bit number with a B_2-bit number gives a $(B_1 + B_2)$-bit product. Because of the recursive nature of the computation in (8.1), the $(B_1 + B_2)$-bit product must be approximated by fewer bits or else the word length would grow without bound. As described in Section 5.1, either truncation or rounding may be used to give a B-bit approximation to the $(B_1 + B_2)$-bit number. The difference between the true product $z = x \cdot y$ and the approximate B-bit representation $[z]_Q$, $e = z - [z]_Q$, is modeled as a uniformly distributed random variable that is independent of the value of z. As shown in Section 5.1, the variance of this quantization noise is $Q^2/12$, where the quantization step size $Q = 2^{-B+1}$. When rounding is used, the noise n is called *roundoff noise* and has zero mean. Rounding will be assumed for the remainder of this discussion. When the product is rounded to B bits, the noise has, from (5.10), a variance of $2^{-2B}/3$.

For the purpose of roundoff noise analysis, the digital filter is modeled as a linear, time-invariant system. As in Chapter 5, a noise source with mean zero and variance given by (5.10) is used to represent the rounding error made after multiplication. The noise samples are assumed to be independent, resulting in a white-noise source with a noise power of $2^{-2B}/3$.

The noise power at the output is found by assuming that each noise source is independent of all the others so that the total power is simply the sum of the individual noise powers. The noise power at the output that results from one noise source n_i is found by first calculating the transfer function from the location of the ith noise source, $H_i(z)$, and then evaluating the power by integrating the noise power spectral density[3] to give

$$P_i = \frac{2^{-2B}}{3} \int_{-1/2}^{1/2} |H_i(e^{j2\pi F})|^2 \, dF. \tag{8.36}$$

Figure 8.11 illustrates how the noise sources contribute to the total output noise. The blocks labeled H_i correspond transfer functions from the location of the ith noise source to the output of the filter. Roundoff noise analysis will only be

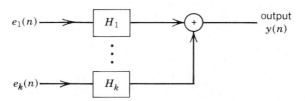

FIGURE 8.11 Contributions of Noise Sources.

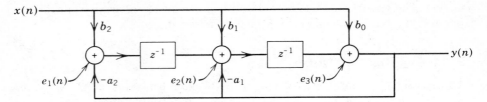

FIGURE 8.12 Transpose Structure with Quantization Noise.

carried out in detail for second-order sections. The principles illustrated by these structures also apply to higher-order sections.

Example 8.3 Noise Power Calculation for a Second-Order Block
    The transpose structure in Fig. 8.3 is reproduced in Fig. 8.12 with the additive quantization noise indicated by additive noise errors $e_i(n)$ at the three places where the signal must be quantized. The transfer function from the first noise source e_1 to the output is

$$\frac{Y(z)}{E_1(z)} = T_1(z) = \frac{z^{-2}}{1 + a_1 z^{-1} + a_2 z^{-2}}. \tag{8.37}$$

The transfer function from the second noise source e_2 to the output is

$$\frac{Y(z)}{E_2(z)} = T_2(z) = \frac{z^{-1}}{1 + a_1 z^{-1} + a_2 z^{-2}}. \tag{8.38}$$

Finally, the transfer function from the third noise source to the output is

$$\frac{Y(z)}{E_3(z)} = T_3(z) = \frac{1}{1 + a_1 z^{-1} + a_2 z^{-2}}. \tag{8.39}$$

All three of these transfer functions have the same squared magnitude, so the noise gain factor for all three is

$$R = \frac{1}{j2\pi} \oint T(z)T(z^{-1}) \frac{dz}{z}. \tag{8.40}$$

After evaluating the integral in (8.40), we get

$$R = \frac{1 + a_2}{(1 - a_2)[(1 + a_2)^2 - a_1^2]}. \tag{8.41}$$

The largest noise gain R occurs when the filter has a double pole near either $+1$

(zero frequency) or -1 (one half of the sampling frequency) with $a_1^2 = 4a_2$. In this case

$$R = R_{max} = \frac{1 + a_2}{(1 - a_2)^3}. \tag{8.42}$$

The smallest noise gain occurs when the filter has poles near plus or minus one quarter of the sampling frequency with $a_1 = 0$ and

$$R = R_{min} = \frac{1}{1 - a_2^2}. \tag{8.43}$$

The total noise power is

$$P = \sum_{i=1}^{3} \frac{R_i Q^2}{12} = 3 \frac{RQ^2}{12}, \tag{8.44}$$

where $Q = 2^{-2B}/3$.

This example shows that the largest noise gain occurs for narrow-band low-pass (poles near $+1$) or high-pass filters (poles near -1). Further, according to (8.42), when the poles are near the unit circle (a_2 close to 1), the noise gain is especially large. The smallest noise gain occurs for filters with the pass band near one half the sampling frequency; poles not too near the unit circle correspond to a small value of the coefficient a_2 (see (8.43)).

Summary

Quantization noise is modeled as independent white-noise sources inserted at each point where the signal is quantized. The contribution of each noise source to the output is determined by the transfer function from the location of the noise source to the output. An example using a transposed second-order section is given. The total noise power at the output of the filter is calculated as the sum of the individual contributions. The transposed structure does not have the best noise characteristics. The direct implementation in Figure 8.1 is better and minimum noise structures are better.

8.2.4 Limit Cycles

In the analysis of recursive filters, we have thus far assumed that the filter was a linear system. The quantization noise analysis in Section 8.3.3 modeled the quantization error as an additive noise source and used linear system theory to provide estimates of the noise power resulting from quantization.

Digital filters are *not* linear systems because of the overflow and quantization phenomena. The overflow phenomenon is a distinctly nonlinear type of behavior. The methods used to handle overflow (two's complement and limiting types) determine the specific type of nonlinearity and the filter's response after an overflow. A digital filter that is stable according to a linear model (all poles inside the unit circle) may nevertheless begin to oscillate when an overflow occurs. This type of oscillation is called a *limit cycle*.[1,10] Example 8.4 illustrates this possibility.

Example 8.4 Two's Complement Limit Cycle
   In this example a second-order filter is shown to exhibit an overflow limit cycle. The filter's transfer function is

$$H(z) = \frac{1}{z^2 - z + \frac{1}{2}}. \tag{8.45}$$

The structure is shown in Fig. 8.13. The block labeled NL represents the nonlinearity that results from two's complement arithmetic. The nonlinear characteristic is illustrated in Fig. 8.14. If the function NL were a linear function, the system would be stable, with poles at $z_{1,2} = 0.5 \pm j0.5$.
   A state-variable analysis of the filter uses the outputs of the delay elements as state variables, as shown in Fig. 8.13.

$$x_1(n + 1) = x_2(n),$$
$$x_2(n + 1) = \text{NL}[-0.5x_1(n) + x_2(n)]. \tag{8.46}$$

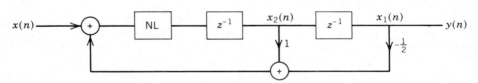

FIGURE 8.13 Direct Structure with Limit Cycles.

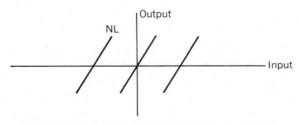

FIGURE 8.14 Two's Complement Overflow Nonlinearity.

With an initial state of $x_1(0) = 0.8$ and $x_2(0) = -0.8$, (8.46) gives

$$x_1(1) = -0.8,$$
$$x_2(1) = \text{NL}[-1.2] = +0.8, \tag{8.47}$$

and for $n \geqslant 1$

$$x_1(n) = (-1)^n 0.8,$$
$$x_2(n) = -(-1)^n 0.8. \tag{8.48}$$

Thus, the system oscillates back and forth between the two states

$$\begin{matrix} x_1 = +0.8, \\ x_2 = -0.8, \end{matrix} \quad \text{and} \quad \begin{matrix} x_1 = -0.8, \\ x_2 = +0.8, \end{matrix} \tag{8.49}$$

and is said to be in a limit cycle.

Overflow limit cycles will not occur in the structure of Fig. 8.13 if there is no overflow to start them and the initial state does not start one. There will be no overflow if the argument of NL is less than 1. In other words,

$$|-a_2 x_1(n) - a_1 x_2(n)| < 1. \tag{8.50}$$

Since $|x_{1,2}| \leqslant 1$, no limit cycles will occur if

$$|a_1| + |a_2| < 1. \tag{8.51}$$

This is a rather severe limitation on the filter coefficients and rules out most practical filters.

Overflow limit cycles can be eliminated by using another nonlinearity, such as the limiting type of nonlinearity shown in Fig. 8.15.

It has been shown[11] that the use of the nonlinearity in Fig. 8.15 will guarantee the absense of large-scale, overflow limit cycles in the direct structure of Fig. 8.2 complement nonlinearity in Example 8.4 is replaced by the nonlinearity in Fig. 8.15, the state decays to zero with zero input. This suggests using the

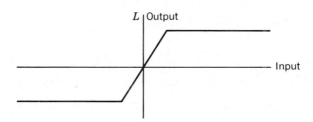

FIGURE 8.15 Limiting Type of Nonlinearity.

nonlinearity of Fig. 8.15, which can be implemented on the TMS32010 chip by setting the overflow mode (OVM). However, it *is* possible to have overflow limit cycles in the direct structure with the nonlinearity of Fig. 8.15 when the input is nonzero, as shown by Example 8.5.

Example 8.5 Limit Cycle with Limiting-Type Nonlinearity
   The filter in this example is the same as in Example 8.4, except that the limiting type of nonlinearity, L, shown in Fig. 8.15, is used. The input is a constant value; that is, $x(n) = -0.5$ for all n. With the limiting type of nonlinearity, L, the state equations are

$$x_1(n + 1) = x_2(n),$$
$$x_2(n + 1) = L[-0.5x_1(n) + x_2(n) - 0.5]. \tag{8.52}$$

With the same initial state as Example 8.4, $x_1(0) = 0.8$ and $x_2(0) = -0.8$, (8.52) gives

$$x_1(1) = -0.8, \qquad x_2(1) = L[-1.2] = -1.0, \tag{8.53}$$

and, for $n \geqslant 2$,

$$x_1(n) = -1.0, \qquad x_2(n) = -1.0. \tag{8.54}$$

The state remains at the constant value of (8.54). This condition is also called a *limit cycle*.
   The state-variable representation of a second-order digital filter can be used to obtain conditions for the absence of limit cycles.[1] A general, linear, second-order system can be expressed in state-variable form by the equations

$$\begin{bmatrix} x_1(n+1) \\ x_2(n+1) \end{bmatrix} = \begin{bmatrix} a_{11} & a_{12} \\ a_{21} & a_{22} \end{bmatrix} \begin{bmatrix} x_1(n) \\ x_2(n) \end{bmatrix} + \begin{bmatrix} b_1 \\ b_2 \end{bmatrix} u(n),$$

$$y(n) = \lfloor c_1 \quad c_2 \rfloor \begin{bmatrix} x_1(n) \\ x_2(n) \end{bmatrix} + [d]u(n). \tag{8.55}$$

If the system is stable (both eigenvalues of the A matrix in (8.55) are <1) and if the quantization corresponds to a nonlinearity with the property that

$$[|x|]_Q \leqslant |x|, \tag{8.56}$$

The system is stable *if and only if* either of the two conditions holds[1]:

(a) $a_{12}a_{21} \geqslant 0,$
(b) $a_{12}a_{21} < 0$ but $|a_{11} - a_{22}| + \det A < 1. \tag{8.57}$

Both the two's complement and the limiting nonlinearities satisfy condition (8.56). For example, the direct structure corresponds to

$$A = \begin{bmatrix} 0 & 1 \\ -a_2 & -a_1 \end{bmatrix}. \tag{8.58}$$

From condition (b) of (8.57), we get

$$|a_{11} - a_{22}| + \det A = |a_1| + a_2 < 1 \tag{8.59}$$

as the necessary and sufficient conditions for absence of limit cycles in the direct structure.

As another example, consider the A matrix for the coupled-form structure, which is stable for $r < 1$.

$$A = \begin{bmatrix} r\cos(\theta) & -r\sin(\theta) \\ r\sin(\theta) & r\cos(\theta) \end{bmatrix}. \tag{8.60}$$

For $\sin(\theta) \neq 0$, $a_{12}a_{21} < 0$, and condition (b) of (8.57) gives

$$|a_{11} - a_{22}| + \det A = 0 + r^2(\cos^2(\theta) + \sin^2(\theta)) < 1. \tag{8.61}$$

From (8.61) we see that the coupled-form structure is free of overflow limit cycles for both the two's complement and the limiting-type of nonlinearities when the input is zero.

If the nonlinearity is limiting type, then it has been shown[4] that no overflow limit cycles with a nonzero input exist when the conditions in (8.57) are satisfied. A complete analysis of limit cycles, both overflow and small scale, is contained in reference 4. Conditions for stability are given in terms of allowed coefficient ranges for several types of overflow nonlinearities and for direct, coupled, wave digital, and lattice structures.

Small-Scale Limit Cycles
The overflow limit cycles have full-scale amplitude and can overwhelm any signal components. The conditions for eliminating this type of limit cycle depend on the filter structure and the way that overflow is handled. There is another type of limit cycle that has a much smaller amplitude and depends on the type of quantization used after a multiplication and on the structure of the filter.[10]

Small-scale limit cycles often occur when the input to the filter is a constant and products are rounded. The rounding itself introduces small amplitude oscillations in the filter. An estimate of the amplitude of the limit cycles has been given by Jackson[2] for a second-order block with a denominator

$$A(z) = z^2 + a_1 z + a_2. \tag{8.62}$$

For a B-bit word length, when rounding is used, the maximum magnitude of a small-scale limit cycle is estimated to be

$$M = 2^{-B+1} \left| \frac{0.5}{1 - |a_2|} \right|_{\text{INT}}, \tag{8.63}$$

where x_{INT} means the smallest integer less than or equal to x. Equation (8.63) implies that the amplitude of small-scale limit cycles can be reduced by increasing the word length (increasing B) and/or by reducing the magnitude of a_2. Reducing the size of a_2 corresponds to moving the poles away from the unit circle.

Truncation, rather than rounding, is recommended to eliminate small-scale limit cycles. For example, the coupled-form structure will have small-scale limit cycles with rounding but will not have small-scale limit cycles when truncation is used. See references 2 and 4 for more detail on small-scale limit cycles.

Summary

Section 8.2 covered finite word-length effects for recursive filters. Coefficient quantization was shown to limit the possible pole locations and therefore limit the possible frequencies of an oscillator. The degradation of frequency response of a filter due to coefficient quantization was corrected by redesigning the unquantized coefficient filter with possibly higher order. The scaling and quantization noise problems were evaluated in detail for second-order sections. Filters with poles near the unit circle were shown to have more serious quantization noise problems.

Limit-cycle oscillations were shown to result from the nonlinearities inherent in a digital filter implementation. The possibility of overflow limit cycles of large amplitude could be reduced by using limiting-type overflow characteristics. Small-scale limit cycles were shown to have an amplitude that could be reduced by using more bits, by moving poles away from the unit circle, or by using truncation arithmetic.

8.3 MINIMUM-NOISE FILTER REALIZATIONS

As discussed in Section 8.1, many different filter structures have the same transfer function. One structure is obtained from another by use of the transformation matrix M. Mullis and Roberts[12] have shown how to transform the state representation of a filter to obtain the minimum possible quantization noise. Although the minimum-noise structure can be derived for any order filter, the number of multiplications proportional to N^2 for an Nth-order filter, becomes prohibitive for high-order filters. A compromise realization uses second-order blocks that individually have the minimum-noise structure in a parallel or cascade connection. The overall structure will *not* have the minimum possible noise, but it will have low quantization noise, low sensitivity to coefficient variations, and a reasonably low number of multiplications.

A derivation of the results of Mullis and Roberts[12] is not presented here. Instead, we give the equations for a second-order minimum-noise structure[9]. This second-order section can then be used in parallel or cascade connections for higher-order filters, as described in Section 8.1.

For a transfer function,

$$H(z) = d + \frac{q_2 z + q_1}{z^2 + p_1 z + p_2},\qquad(8.64)$$

the direct form has the state-variable representation

$$A = \begin{bmatrix} 0 & 1 \\ -p_2 & -p_1 \end{bmatrix}, \quad B = \begin{bmatrix} 0 \\ 1 \end{bmatrix}, \quad C = |q_1 \quad q_2|, \quad D = d, \quad (8.65)$$

and the minimum-noise structure has

$$A' = \begin{bmatrix} R & -\dfrac{Id_2}{d_1} \\ \dfrac{Id_1}{d_2} & R \end{bmatrix}, \quad B' = \begin{bmatrix} \dfrac{\sin(\phi/2)}{d_1} \\ \dfrac{\cos(\phi/2)}{d_2} \end{bmatrix}, \qquad (8.66)$$

$$C' = r \left| d_1 \cos\left(\frac{\phi}{2}\right) \quad d_2 \sin\left(\frac{\phi}{2}\right) \right|, \quad D' = [d].$$

The parameters in the minimum-noise representation (8.66) for poles at $R \pm jI$ are

$$r^2 = q_2^2 + \left(\frac{q_1 + Rq_2}{I}\right)^2,$$

$$\phi = \tan^{-1}\left(\frac{-Iq_2}{q_1 + Rq_2}\right),$$

where d_1 and d_2 are scaling constants based on appropriate norms of the response of the first and second state variables to a unit pulse input. (See Section 8.2.2).

Since the minimum-noise filter has $a_{12}a_{21} < 0$, $a_{11} = a_{22}$, and $\det A < 1$, the minimum-noise filters satisfy the conditions (8.57) for stability. If the limiting-type of nonlinearity shown in Fig. 8.15 represents the way that overflow is treated, then the minimum-noise filter will not have overflow limit cycles regardless of whether the input is zero or not.

8.4 DESIGN EXAMPLE

This design example gives a detailed five-step design and implementation of a fourth-order elliptic filter. The cascade of two second-order blocks is used. Each block is implemented in the transpose structure. The poles are paired with the closest zeros. The section with poles nearest the unit circle is used at the output. Scaling for the filter is performed first for the first second-order section. The impulse response of this scaled first section is then convolved with each of the appropriate impulse responses of the second section, and scaling is done on the second section.

STEP 1. The first step in the design is to decide on the filter specifications. For this example the specifications call for a fourth-order elliptic filter. The specifications and the output of Program 9 are given in Fig. 8.16.

a)
Desired pass-band edge 0.25
Desired stop-band edge 0.30
Desired pass-band max. attn. 0.5 dB
Desired stop-band min. attn. 32 dB

b)

$$H(z) = 0.1473035 \cdot \frac{1 + 1.621784z^{-1} + z^{-2}}{1 - 0.4030703z^{-1} + 0.2332662z^{-2}} \cdot \frac{1 + 0.7158956z^{-1} + z^{-2}}{1 + 0.0514214z^{-1} + 0.7972861z^{-2}}$$

c)

Real part	Imaginary part	Magnitude	Phase
	Zeros		
$-0.8108920e+00$	$0.5851958e+00$	$0.1000000e+01$	$0.2516471e+01$
$-0.8108920e+00$	$-0.5851958e+00$	$0.1000000e+01$	$-0.2516471e+01$
$-0.3579478e+00$	$0.9337416e+00$	$0.1000000e+01$	$0.1936865e+01$
$-0.3579478e+00$	$-0.9337416e+00$	$0.1000000e+01$	$-0.1936865e+01$
	Poles		
$0.2015399e+00$	$0.4389205e+00$	$0.4829798e+00$	$0.1140341e+01$
$0.2015399e+00$	$-0.4389205e+00$	$0.4829798e+00$	$-0.1140341e+01$
$-0.2570916e-01$	$0.8925394e+00$	$0.8929096e+00$	$0.1599593e+01$
$-0.2570916e-01$	$-0.8925394e+00$	$0.8929096e+00$	$-0.1599593e+01$

FIGURE 8.16 Fourth-order elliptic low-pass design example (a) specifications; (b) Transfer function; (c) zeros and poles.

STEP 2. The next step is to decide on the structure for implementing the filter, as described in Section 8.1. The cascade structure in Fig. 8.5 was chosen for this example. Each second-order block was implemented with the transpose structure shown in Fig. 8.3. The poles farthest from the unit circle were used for the first section in Fig. 8.17. The pair of zeros closest to these poles was used.

STEP 3. To scale section 1, we calculated the impulse response and frequency response from the input x to each of three points where overflow could occur; these points are labeled y_{11}, y_{12}, and y_1 in Fig. 8.17a. For the impulse response to the output of the section, y_1, the l_1 and l_2 norms are 5.30748 and 2.7843, respectively. The maximum value of the frequency response, shown in Fig. 8.18,

$$A(11) = -0.4030702997$$
$$A(12) = 0.2332661953$$
$$B(10) = 1.0$$
$$B(11) = 1.621783996$$
$$B(12) = 1.0$$

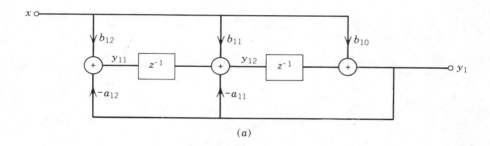

(a)

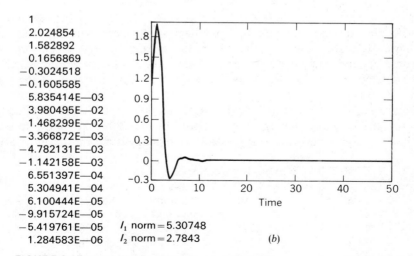

```
1
2.024854
1.582892
0.1656869
-0.3024518
-0.1605585
5.835414E—03
3.980495E—02
1.468299E—02
-3.366872E—03
-4.782131E—03
-1.142158E—03
6.551397E—04
5.304941E—04
6.100444E—05
-9.915724E—05
-5.419761E—05
1.284583E—06
```

l_1 norm $= 5.30748$
l_2 norm $= 2.7843$ (b)

FIGURE 8.17 (a) Coefficients for Section 1; (b) Impulse response to y_1 for Section 1.

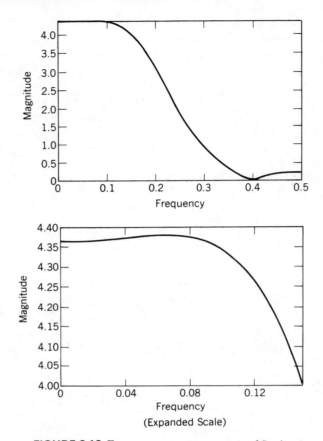

FIGURE 8.18 Frequency response to output of Section 1.

is 4.38. The impulse response y_{11} and its Fourier transform, the frequency response H_{11}, are shown in Figs. 8.19 and 8.20, respectively. The impulse response y_{12} and frequency response H_{12} are shown in Figs. 8.21 and 8.22. The three different measures of gain are shown in Table 8.1 for each of the three locations in Section 1.

Figure 8.23 shows the scaled coefficients for section 1, which were obtained by dividing the original numerator coefficients by the l_1 norm of the impulse response y_1, the largest l_1 norm. This scaling strategy is the most conservative.

TABLE 8.1. Norms for Scaling Section 1

location	l_1 norm	l_2 norm	max $|H(f)|$
1	5.30748	2.7843	4.38
11	1.7715	0.9774	1.28
12	4.30748	2.5985	3.67

Transfer function to y_{11}:

$$H_{11}(z) = \frac{Y_{11}}{X} = \frac{b_{110} + b_{111}z^{-1}}{1 + a_{111}z^{-1} + a_{112}z^{-2}}$$

$A_{(111)} = -0.4030702997$
$A_{(112)} = 0.2332661953$
$B_{(110)} = 0.766733805$
$B_{(111)} = -0.781377681$
$B_{(112)} = 0.0$

(a)

0.7667338
−0.4723301
−0.3692353
−3.864915E—02
7.055178E—02
3.745287E—02
−1.361206E—03
−9.28515E—03
−3.425045E—03
7.853776E—04
1.11551E—03
2.664268E—04
−1.528219E—04
−1.237464E—04
−1.423029E—05
2.313003E—05
1.264247E—05
−2.99649E—07

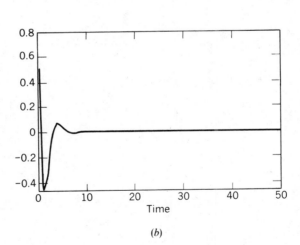

(b)

I_1 norm $= 1.7715$
I_2 norm $= 0.9774$

FIGURE 8.19 (a) Transfer function to location y_{11} in Section 1; (b) Impulse response to location y_{11} in Section 1.

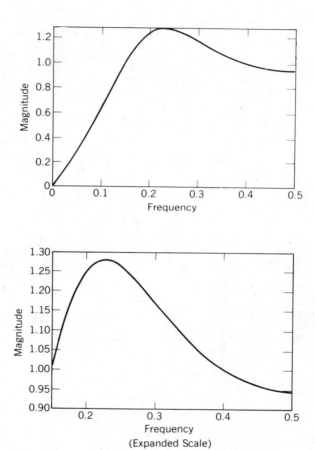

FIGURE 8.20 Frequency response to location y_{11} in Section 1.

Section 1   Transfer function to y_{12}:

$$H_{12} = \frac{Y_{12}}{X} = \frac{b_{120} + b_{121\,121}z^{-1}}{1 + a_{121}z^{-1} + a_{122}z^{-2}}$$

$A_{(121)} = -0.4030702997$
$A_{(122)} = 0.2332661953$
$B_{(120)} = 2.024854296$
$B_{(121)} = 0.766733805$
$B_{(121)} = 0.766733805$
$B_{(122)} = 0.0$

(a)

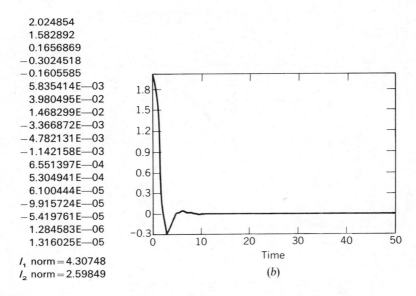

2.024854
1.582892
0.1656869
−0.3024518
−0.1605585
5.835414E—03
3.980495E—02
1.468299E—02
−3.366872E—03
−4.782131E—03
−1.142158E—03
6.551397E—04
5.304941E—04
6.100444E—05
−9.915724E—05
−5.419761E—05
1.284583E—06
1.316025E—05

l_1 norm = 4.30748
l_2 norm = 2.59849

(b)

FIGURE 8.21 (a) Transfer function to location y_{12} in Section 1; (b) Impulse response to location y_{12} in Section 1.

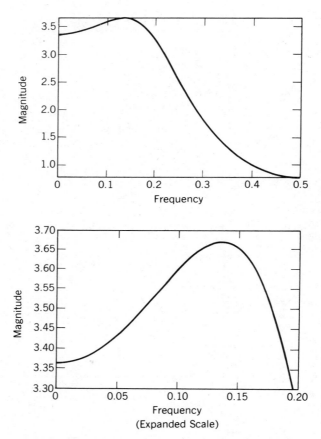

FIGURE 8.22 Frequency response to location y_{12} in Section 1.

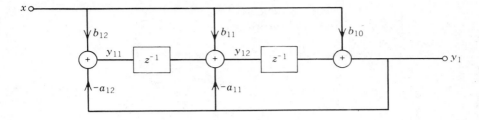

$$H_1(z) = \frac{b_{10} + b_{11}z^{-1} + b_{12}z^{-2}}{1 + a_{11}z^{-1} + a_{12}z^{-2}}$$

Scaled, quantized coefficients:

(original *b* coeff. divided by 5.30748)

	decimal	hex
$a_{11} =$	−0.4030703	CC68
$a_{12} =$	0.2332662	1DDC
$b_{10} =$	0.1884133	181E
$b_{11} =$	0.3055656	271D
$b_{12} =$	0.1884133	181E

FIGURE 8.23 Scaled, quantized coefficients for Section 1.

STEP 4. To scale section 2, we calculated the impulse response and frequency response from the second stage input, x_2 to each of three points where overflow could occur; these points are labeled y_{21}, y_{22}, and y_2 in Fig. 8.24a. Because we are interested in scaling according to the input, x, of the filter, not the input of the second section, these three impulse responses are convolved with the impulse response of the scaled section 1. In this way the impulse response is calculated from the filter input x to the three locations in the second section, y_{21}, y_{22}, and

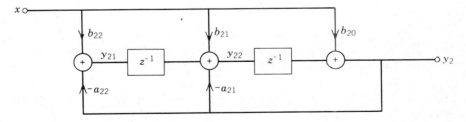

A(21) = 0.05142140356
A(22) = 0.7972860789
B(20) = 1.0
B(21) = 0.7158955928
B(22) = 1.0

FIGURE 8.24 Coefficients for Section 2.

TABLE 8.2. Norms for Scaling Section 2

location	l_1 norm
2	2.85527
21	1.36112
22	1.86357

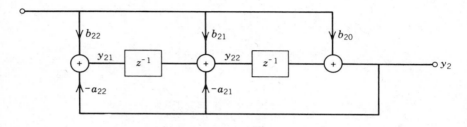

$$H_2(z) = \frac{b_{20} + b_{21}z^{-1} + b_{22}z^{-2}}{1 + a_{21}z^{-1} + a_{22}z^{-2}}$$

scaled, quantized coefficients:

(original *b* coefficients divided by 2.855274)

Decimal	Hex
$a_{21} = 0.0514214$	0695
$a_{22} = 0.7972861$	660D
$b_{20} = 0.35022908$	1A00
$b_{21} = 0.25072745$	129D
$b_{22} = 0.35022908$	1A00

FIGURE 8.25 Scaled, quantized coefficients for Section 2.

y_2. For the impulse response to the output of the section, y_2, the l_1 norm is 2.85527. The other two impulse response responses have smaller norms, as shown in Table 8.2.

Figure 8.25 shows the scaled coefficients for section 2, which were obtained by dividing the original numerator coefficients by the l_1 norm of the impulse response y_2, the largest l_1 norm. This scaling strategy is the most conservative.

STEP 5. The scaled coefficients calculated in steps 3 and 4 were used in an assembly language program implementing the cascade of two transpose structure second-order sections. Program 12 in the appendix is a complete assembly language program for the TMS32010.

The entire five-step procedure was repeated for a cascade of two second-order sections where each section was implemented in the direct form. The details have been omitted since they are essentially the same as for the transpose structures. Program 13 is an assembly language program for this direct structure implementation.

REFERENCES

[1]   R. A. Roberts and C. T. Mullis, *Digital Signal Processing*, Reading, MA: Addison-Wesley, 1987.

[2]   L. B. Jackson, *Digital Filters and Signal Processing*, Boston: Kluwer, 1986.

[3]   L. R. Rabiner and B. Gold, *Theory and Application of Digital Signal Processing*, Englewood Cliffs, NJ: Prentice-Hall, 1975.

[4]   K. T. Erickson and A. N. Michel, "Stability Analysis of Fixed-Point Digital Filters Using Generated Lyapunov Functions—Parts I and II," *IEEE Trans. Circuits Systems* **CAS-32**, 113–142 (1985).

[5]   J. F. Kaiser, "Some Practical Considerations in the Realization of Linear Digital Filters," *Proceedings of the Third Allerton Conference on Circuit and System Theory*, pp. 621–633, October 1965.

[6]   C. M. Rader and B. Gold, "Effects of Parameter Quantization on the Poles of a Digital Filter," *Proc. IEEE* **55**, 688–689 (1967).

[7]   A. H. Gray and J. D. Markel, "Digital Lattice and Ladder Filter Synthesis," *IEEE Trans. Audio Electroacoustics* **AU-21**, 491–500 (1973).

[8]   A. Fettweis, "Digital Filter Structures Related to Classical Filter Networks," *Arch. Elek. Ubertragung* **25**, 79–89 (1971).

[11] P. M. Ebert, J. E. Mazo, and M. G. Taylor, "Overflow Oscillations in Digital Filters," *Bell System Tech. J.* **48**, 2999–3020 (1969).

[9]   H. W. Schüssler, *Notes for a Seminar on Wordlength Effects in Nonrecursive and Recursive Filters*, Rice University, 1984.

[10] H. W. Schüssler, *Digitale Systeme zur Signalverarbeitung*, New York: Springer-Verlag. 1973.

[11] P. M. Ebert, J. E. Mazo, and M. G. Taylor, "Overflow Oscillations in Digital Filters," *Bell System Tech. J.* **48**, 2999–3020 (1969).

[12] C. T. Mullis and R. A. Roberts, "Synthesis of Minimum Roundoff Noise Fixed Point Digital Filters," *IEEE Trans. Circuits Systems* **CAS-23**, 551–561 (1976).

Part IV

Summary

9

Summary

This summary chapter reviews the highlights of the book, comparing and relating the various aspects of the approximation and realization problems in digital filter design. Section 9.1.1 summarizes the key features of FIR filters, and Section 9.1.2 focuses on IIR filters.

9.1 COMPARISON OF FILTERING ALTERNATIVES

9.1.1 FIR Digital Filters

An FIR digital filter has a finite-duration unit-pulse response. Its transfer function is a polynomial in z^{-1}.

$$H(z) = h_0 + h_1 z^{-1} + \cdots + h_{N-1} z^{-(N-1)}. \tag{9.1}$$

A length-N filter has a transfer function that has $N - 1$ zeros in the z plane and has an order $N - 1$ pole at the origin of the z plane.

$$H(z) = \frac{h_0 z^{N-1} + \cdots + h_{N-1}}{z^{N-1}}. \tag{9.2}$$

An FIR filter is called an *all-zero filter* because it has zeros but no poles other than that at the origin.

An FIR digital filter can have exactly linear phase. In other words, the group delay of the filter can be a constant. This linear-phase property results from symmetry of the unit-pulse response of the filter. An IIR filter has an infinite-duration unit-pulse response that cannot be symmetric if it is causal ($=0$ for $n < 0$). Therefore, an IIR filter cannot have exactly linear phase. Of course, an

269

IIR filter can be designed with a good approximation to linear phase, at least over a limited band of frequencies. The delay of a causal linear-phase FIR filter of length N is exactly $(N - 1)/2$. The required filter length N increases when sharp transitions between frequency bands are specified, and/or large attenuations are required in the stop bands. Thus, high-performance, linear-phase filters with sharp cutoffs and large attenuations necessarily are long and have large, though constant, delays and a large number of coefficients to be stored.

When a precisely constant delay is not required for all frequencies, better FIR filters can be designed. When the group delay is of little concern, the minimum-phase FIR filter may be a good choice. We can design a filter that has the best magnitude characteristics, in the Chebyshev sense, and a minimum phase shift. These minimum-phase filters generally have better magnitude characteristics for the same length N than do linear-phase filters. The group delay, though minimal, is usually far from a constant. For low-pass filters the delay is quite small at zero frequency and increases rapidly near the band edge. When a better group delay is required, complex approximation techniques can be applied to give a good, small, though not exactly constant, group delay with good magnitude characteristics.

One new problem with the complex design of FIR filters is the specification of the desired group delay or phase. If too small a delay is requested, the best Chebyshev approximation has large errors and the filter is not useful. Generally, a delay between one half and three quarters of the delay of the same length linear-phase filter is a good choice. The choice of desired delay depends, of course, on the band edges specified for the filter. Wide-band filters can have less delay, for the same Chebyshev error, than narrow-band filters can.

The approximation problem for both FIR and IIR filters is solved in one of two ways. Either a closed-form, analytic expression is used, possibly with suitable transformations, or a numerical optimization procedure is used to solve for the coefficients of the filter. One family of closed-form analytical design formulas for FIR filters gives an optimal LS error approximation to an ideal low-pass filter with a spline or trigonometric function transition region. A second family is based on windowing the design of a LS error approximation in order to reduce the Chebyshev error at the expense of the squared error.

Numerical procedures for FIR design may be divided into two categories. Methods like frequency sampling and LS error minimization require solving a set of linear equations. Other methods, such as linear programming and the Remes exchange algorithm, are iterative and generally take more time than the frequency-sampling and LS methods.

Programs are provided in the appendix to design FIR filters using windowing, frequency sampling, and LS-error minimization. A program is also provided for the Parks–McClellan algorithm, which designs filters with minimum Chebyshev error and equiripple frequency characteristic. It usually takes longer to design an FIR filter with an iterative procedure like the Parks–McClellan algorithm than it does to design an IIR filter of approximately equivalent performance with analytic expressions and transformations. However, a wider range of specifications can be met with the numerical optimization approach than with the analytical approach to the approximation problem. As described in Chapter 4, FIR filters can be designed to meet arbitrary complex-frequency specifications with minimal Chebyshev error by using linear programming.

The realization or implementation of an FIR filter with fixed-point arithmetic is much easier and more trouble free than the implementation of an IIR filter. The direct, nonrecursive implementation of the FIR filter where the output is calculated as a weighted linear combination of present and past inputs is always stable. For filter lengths of up to 100 the coefficients are not very sensitive to quantization. The unit-pulse response coefficients can be implemented with 12 to 16 bits with little degradation of the frequency response. If shorter word length is desired, an optimization program is available that will solve the approximation problem with quantized coefficients. The coefficients of the filter are easily scaled to avoid overflow, and the quantization noise problems are not severe when the products of filter coefficients and input samples are accumulated in a double word-length register. The only significant difficulty in implementing FIR filters is the large amount of memory required to store the present and past N values of the input signal for a length-N filter. The FIR filter also has more filter coefficients to store than an IIR filter with similar performance. This problem should become less important as memory becomes more readily available. Another possible disadvantage of the large number of coefficients in the FIR filter arises when the coefficients are changed often, as in adaptive filtering applications.

Long FIR filters can also be efficiently implemented by using special hardware, such as array processors, for computing the required inner products. The necessary convolution can be implemented by fast convolution techniques using the FFT or by table lookup, using distributed arithmetic.

9.1.2 IIR Digital Filters

An IIR filter has an infinite-duration unit-pulse response. The transfer function of an IIR filter, as described in this book, is a rational function of z^{-1}:

$$H(z) = \frac{b_0 + b_1 z^{-1} + \cdots + b_M z^{-M}}{1 + a_1 z^{-1} + \cdots + a_N z^{-N}}, \tag{9.3}$$

This function is also written as a rational function of z:

$$H(z) = z^{N-M} \frac{b_0 z^M + b_1 z^{M-1} + \cdots + b_M}{z^N + a_1 z^{N-1} + \cdots + a_N}. \tag{9.4}$$

Unlike analog filters, where the order of the numerator must be less than or equal to the order of the denominator, a digital filter can have N greater than, equal to, or less than M. In addition to the order $N - M$ zero or pole at the origin, the filter has M zeros and N poles in the z plane.

An IIR filter can generally achieve a sharper transition between band edges than an FIR filter can with the same number of coefficients. The reason is that the IIR filter has a pole near the edge of the pass band and a nearby zero at the edge of stop band. Since an FIR filter cannot have poles (except at the origin), it cannot achieve the same sharp cutoff. The closely spaced pole and zero, which produce the desired sharp change in magnitude characteristic of the filter, also produce a rapid change in phase and phase slope for frequencies near the pole and zero. The closely spaced pole and zero lead to a rapid change in group delay for frequencies approaching the band edge. It is not possible for an IIR filter to have exactly constant group delay for all frequencies. Minimum-phase, low-pass IIR filters with sharp transitions between the pass band and stop band typically have a small group delay at zero frequency that increases rapidly at frequencies near the band edge.

An IIR filter can have precisely constant magnitude (an all-pass filter). All-pass IIR filters can be used as phase or delay equalizers to compensate for the delay distortion present in minimum-phase systems. However, all-pass IIR equalizers are difficult to design. Recent work has shown that FIR equalizers, designed with the techniques described in Chapter 4, can have characteristics similar to, if not better than, IIR equalizers. Furthermore, FIR equalizers are easier to implement than IIR equalizers.

Implementing IIR filters with a recursive realization in fixed-point arithmetic is much more difficult than the direct, nonrecursive implementation of an FIR filter. Much greater care must be taken in the scaling of the filter coefficients. When there is an overflow in a recursive filter, large-scale oscillations (limit cycles) can occur, which obscure any useful output from the filter. Because of internal rounding of the signal variables, small-scale limit cycles can also occur, adding a small but annoying noise to the filter output. The design example in this book has a small-scale limit cycle.

Quantization noise can be more of a problem than in nonrecursive filters because of the recursive nature of the calculation. The double-length product of two numbers must be quantized in order to be fed back in the recursion. The frequency response, and even the stability, as determined from the pole locations, is sensitive to quantization of the filter coefficients. This sensitivity rules against directly implementing the difference equation implied by (9.3). Cascade or parallel connections of low-order blocks are better implementations for recursive filters.

An IIR filter has an advantage over an FIR filter in that it generally has fewer coefficients than an FIR filter with similar magnitude characteristics, so less memory is required to store the coefficients. A more significant memory saving occurs because only a few of the recent input values need to be stored, in contrast to the FIR case where N input values need to be stored for a length-N filter. Even though the IIR filter has an infinite memory—that is, its output depends on the infinite past input—the filter memory is stored in the state variables of the recursive filter. The memory of the filter arises from the storage of past outputs as well as past inputs.

Even though the IIR filter has fewer coefficients than an equivalent FIR filter, it still may take less time to compute an output sample for the equivalent FIR filter. The reason is the regularity of the VLSI structure for implementing the nonrecursive filter as compared to the irregular structure required for the recursive filter. For example, with the TMS32020 signal processor, the nonrecursive calculation requires approximately one fifth of the time per coefficient of the recursive calculation. In other words, for the same computing time the FIR filter can have approximately five times as many coefficients as an IIR filter. The exact relation for computing times depends, of course, on the particular programs used to implement the filters. In applications where the coefficients of the filter are updated in real time (e.g., adaptive filtering), the advantage of fewer coefficients in the IIR filter may be significant. Distributed arithmetic is more attractive for IIR filters than for FIR filters because of the lower order. However, the use of FFTs for implementing an IIR filter requires a block recursive structure and is not as effective as for the FIR filter.

9.2 DESIGN ENVIRONMENT

Because of interrelated steps in the approximation and realization parts of the filter design process, interactive design programs must be available on a computer. The FORTRAN programs in the appendix and/or those available in the IEEE Press program book or from commercial sources can provide that environment for the approximation problem. The realization problem requires a simulation program to analyze the quantization effects in a particular realization of a filter. The simulator must be specialized for the particular hardware or computer implementation; therefore, it is not included in this book. Certain simulation programs are available for the TMS320 family of signal processors from Texas Instruments, Inc., and others. A fairly general program, called DOREDI, for analyzing the effects of finite word-length effects in realizations of IIR filters is available in the IEEE Press program book.

Appendix

This appendix contains FORTRAN programs for designing FIR and IIR filters. Most of the programs are written with a notation and organization that follows the theoretical development in the book. Studying the programs should help you understand the theory, and vice versa. They are written to utilize very efficient algorithms and formulas, but they do not incorporate all the user-friendly input/output characteristics or error-handling capabilities of commercial products. The exception is the Parks–McClellan program, which has been developed over several years. The programs, in general, use a basic structure that the user can modify as necessary.

1. A FORTRAN PROGRAM FOR LINEAR-PHASE LOW-PASS FIR FILTER DESIGN USING FREQUENCY SAMPLING

The FORTRAN program is a system for designing a length-N, linear-phase, low-pass FIR digital filter with a frequency response that interpolates N specified values. These N values are usually samples of a desired continuous-frequency response. After the filter is designed, its frequency response is calculated for analysis. The basic theory, formulas, variable names, and references are chosen to follow the development in Section 3.1.

The main program starts with a section that takes input specifications from the terminal. The length of the desired FIR filter is entered as N. Next follows the cutoff frequency or band edge in hertz, where we assume a sampling rate of 1.0 and N equally spaced frequency samples. The next input distinguishes the two possible sampling schemes described in Section 3.1. Entering 0 for DC specifies a sample at $\omega = 0$, and entering 1 specifies that the samples are shifted one-half interval, so there is no sample at $\omega = 0$. A value for K is entered to set the number of equally spaced frequencies at which the frequency response is evaluated in the analysis section of the program.

In the next section of the program, the desired frequency-response samples of a low-pass filter are loaded into the array $A(J)$. These are set to be 1 or 0, according to the input specification FP. This section would be changed in order to design something other than a simple low-pass filter.

The actual design of the filter is performed in the DO 15 and DO 21 loops, where design formulas (3.4), (3.6), (3.10), and (3.12) are evaluated. The first half of the symmetric impulse response is written to the terminal as the coefficients of the designed filter. The frequency response of the filter is calculated at K equally spaced frequencies by the subroutine FREQ( ), which implements (3.2) and (3.5); these values are written to the file fm. If K is set equal to N, the output of the frequency-response calculation should give the 1's and 0's that were the input samples in the array A(J).

This program was used to design the filters in Examples 3.1 and 3.2, and a modified version was used for Example 3.4. It could have also been used in Example 3.3. The input and A(J)-setting sections could easily be modified to allow any desired samples to be specified. Because this scheme is an interpolation method, the analysis of the frequency over a fairly large range is important for examining the behavior between the sample points.

As pointed out in Section 3.2.1, this frequency-sampling design program can be used to design optimal LS error approximations over L frequency samples by designing a length-L filter and symmetrically truncating the impulse response to the desired length N. It will give the same results as Program 2 for equally spaced samples and no weighting, but it runs faster and has less numerical error.

```
C          FREQUENCY SAMPLING OR INTERPOLATION FOR FIR FILTERS
C          TRUNCATION YIELDS OPTIMAL LEAST-SQUARES DESIGN
C          DESIGN PROGRAM FOR A LINEAR PHASE LOWPASS FILTER
C             FILTER LENGTH AND NO. OF FREQ SAMPLES = N
C             BANDEDGE IN HZ = FP,   FOR SAMPLING RATE = 1
C             FREQ. SAMPLE AT DC: DC = 0
C             NO FREQ. SAMPLE AT DC: DC = 1
C             FREQUENCY RESPONSE CALCULATED AT K POINTS
C          C.S. BURRUS,   RICE UNIVERSITY,   JAN 1987
C--------------------------------------------------------------
           REAL X(101), A(101), B(1001)
C-------------------INPUT-----------------------------
           WRITE (6,100)
      5    WRITE (6,110)
           READ (5,*) N, FP, DC, K
           M   = (N+1)/2
           AM  = (N+1.0)/2.0
           M1  = N/2 + 1
           Q   = 6.28318530717959/N
           N2  = N/2
             NP = N*FP + 1.0
           IF (DC.EQ.0) GOTO 10
             NP = N*FP + 0.5
```

```
C------------SET DESIRED FREQ RESPONSE--------------------
  10      DO 11 J = 1, NP
            A(J) = 1.0
  11      CONTINUE
          DO 12 J = NP+1, M1
            A(J) = 0.0
  12      CONTINUE
          IF (DC.EQ.1) GOTO 18
C------------TYPE 1&2, DC FREQ SAMPLE--------------------
          DO 15 J = 1, M
            XT = A(1)/2.0
            DO 14 I = 2, M
              XT = XT + A(I)*COS(Q*(AM-J)*(I-1))
  14        CONTINUE
            X(J) = 2.0*XT/N
  15      CONTINUE
          GOTO 21
C------------TYPE 1&2, NO DC FREQ SAMPLE-----------------
  18      DO 21 J = 1, M
            XT = 0
            DO 20 I = 1, N2
              XT = XT + A(I)*COS(Q*(AM-J)*(I-0.5))
  20        CONTINUE
            IF (AM.NE.M) XT = XT + A(M)*COS(3.141592654*(AM-J))/2
            X(J) = 2*XT/N
  21      CONTINUE




C-----------------OUTPUT----------------------------------.
          WRITE (6,120)(X(J),J=1,M)
          CALL FREQ(X,B,N,K)
          OPEN (1,FILE ='fm')
          REWIND(1)
          DO 50 J = 1, K+1
            F = 0.5*(J-1)/K
            WRITE (1,130) F, ABS(B(J))
  50      CONTINUE
 100      FORMAT ('FREQUENCY SAMPLING DESIGN OF A LOWPASS FILTER')
 110      FORMAT ('ENTER: N, FP, DC, K')
 120      FORMAT (9F8.5)
 130      FORMAT (5X,F15.8,E15.8)
          GOTO 5
          END
C-----------------END OF MAIN PROGRAM--------------------
          SUBROUTINE FREQ(X,A,N,K)
          REAL X(1), A(1)
C
          Q = 3.141592654/K
          AM = (N+1)*0.5
          M  = (N+1)/2
          N2 = N/2
          DO 20 J = 1, K+1
            AT = 0
            IF (AM.EQ.M) AT = 0.5*X(M)
            DO 10 I = 1, N2
              AT = AT + X(I)*COS(Q*(AM-I)*(J-1))
  10        CONTINUE
            A(J) = 2*AT
  20      CONTINUE
          RETURN
          END
```

2. A FORTRAN PROGRAM FOR LINEAR-PHASE LOW-PASS FIR FILTER DESIGN USING A DISCRETE LEAST SQUARED ERROR CRITERION

The FORTRAN program is a system for designing a length-N, linear-phase, low-pass FIR digital filter with a frequency response that is a LS error approximation to a set of L desired values. These L values are usually samples of a desired continuous-frequency response. If L is equal to N, the approximation can be exact, and this program gives the same results as a frequency-sampling design. After the filter is designed, its frequency response is calculated for analysis. The basic theory, formulas, variable names, and references are chosen to follow the development in Section 3.2.1.

The main program starts with a section that takes input specifications from the terminal. The length of the desired FIR filter is entered as N. The number of frequencies over which to calculate the approximation is entered as L, where $L \geq N$. Then follows FP, the cutoff frequency or band edge in hertz, where we assume a sampling rate of 1.0 and L equally spaced frequency samples. The next input distinguishes the two possible sampling schemes described in Section 3.1. Entering 0 for DC specifies a sample at $\omega = 0$, and entering 1 specifies that the samples are shifted one-half interval, so there is no sample at $\omega = 0$. A value for K is entered to set the number of frequencies at which the frequency response is evaluated in the analysis section of the program.

In the next section the desired frequency response samples of a low-pass filter are loaded into the array A(J). These samples are set to be 1 or 0, according to the input specifications FP, DC, and N being even or odd. This section would be changed in order to design something other than a simple low-pass filter.

The next section calculates the frequency-response matrix F defined in (2.25) by using (3.2) and (3.5). If $L = N$, the result is the same as frequency sampling. If $L > N$, (2.25) is overdetermined and F is rectangular. An approximate solution to these equations is found by solving the normal equations of (3.19) by solving (3.20). However, this program uses the efficient and accurate subroutines SQRDC( ) and SQRSL( ) contained in the matrix software system LINPACK. These subroutines are covered in reference 7. Their output, which is the solution of the normal equations, is the first half of the symmetric impulse response, which is written to the terminal as the coefficients of the designed filter. The frequency response of the filter is calculated at K equally spaced frequencies by the subroutine FREQ( ), which implements (3.2) and (3.5); these values are written to the file fm.

This program provides the basis of a versatile and powerful FIR filter design system. At one extreme, for $L = N$, it designs according to a frequency-sampling criterion. At the other extreme, for $L \gg N$, it gives a good approximation to the true continuous LS error design used in Programs 3, 4, and 5, yet it allows arbitrary ideal specifications. The input section can easily be modified to accept arbitrary input to the array A(J). However, numerical problems may exist for a long filter with $L \gg N$.

The advantage of this program over Program 1 comes from the ability to modify it for unequally spaced frequency points by changing the section that creates the F matrix. This modification allows us better control of the approximation because we can use a denser grid near a discontinuity. We can add a weight function according to (3.22), which can also allow greater control of the approximation. We can define a transition band by using a small weight in the "don't care" transition region, and by using a general expression for the frequency-response matrix F, rather than the one for linear phase, we can generalize the program to do complex approximation.

If equally spaced frequency samples and no weights are satisfactory, do not use this program. Use Program 1 for a frequency-sampling design with filter length L and truncated that result to the desired length N. That will do the same thing this program does, but it will be faster and have less numerical roundoff error.

```
C          DISCRETE LEAST SQUARE ERROR  FIR  FILTER
C              FILTER LENGTH = N
C              NO. OF FREQUENCY SAMPLES = L
C              BANDEDGE IN HZ = FP    FOR SAMPLING RATE = 1
C              FREQ. SAMPLE AT DC: DC = 0
C              NO FREQ. SAMPLE AT DC: DC = 1
C              FREQUENCY RESPONSE CALCULATED AT K POINTS
C          C.S. BURRUS,    RICE UNIVERSITY,   JAN 1987
C-----------------------------------------------------------------
           REAL X(101), A(501), B(1001)
           REAL F(501,101), QAX(101)
C-------------SET PARAMETERS----------------------------------
           LDX = 501
           WRITE (6,100)
      5    WRITE (6,110)
           READ (5,*) N, L, FP, DC, K
           M  = (N+1)/2
           AM = (N+1)/2.0
           LM = (L+1)/2
             L2 = L/2 + 1
             LP = L*FP + 1.0
             QJ = 1.0
           IF (DC.EQ.0) GOTO 18
             L2 = LM
             LP = L*FP + 0.5
             QJ = 0.5
C-------------SET THE DESIRED FREQ RESPONSE-------------------
           DO 6 J = 1, L
             A(J) = 0.0
      6    CONTINUE
           IF (MOD(N,2).EQ.0) GOTO 8
           DO 7 J = 1, LP
             A(J) = 1.0
             A(L-J+1) = 1.0
      7    CONTINUE
           GOTO 15
      8    DO 9 J = 1, LP
             A(J) = 1.0
             A(L-J+1) =-1.0
      9    CONTINUE
           GOTO 15
```

```
18      DO 10 J = 1, L
           A(J) = 0.0
10      CONTINUE
        IF (MOD(N,2).EQ.0) GOTO 12
        A(1) = 1.0
        DO 11 J = 2, LP
           A(J)     = 1.0
           A(L-J+2) = 1.0
11      CONTINUE
        GOTO 15
12      A(1) = 1.0
        DO 15 J = 2, LP
           A(J)     = 1.0
           A(L-J+2) =-1.0
15      CONTINUE




C-------------CREATE THE FREQ RESPONSE MATRIX: F ---------------
        Q = 6.283185307179586/L
        DO 30 I = 1, M
           QI = Q*(AM-I)
           DO 20 J = 1, L
              F(J,I) = 2*COS(QI*(J-QJ))
20         CONTINUE
30      CONTINUE
C-------------LINPACK LEAST SQUARES SUBROUTINE-----------------
        CALL SQRDC(F,LDX,L,M,QAX,DUM,DUM,0)
        CALL SQRSL(F,LDX,L,M,QAX,A,DUM,A,X,DUM,DUM,100,INFO)
        IF (MOD(N,2).NE.0) X(M) = 2.0*X(M)
C--------------------OUTPUT-------------------------------------
        WRITE (6,120)(X(J),J=1,M)
        CALL FREQ(X,B,N,K)
        OPEN (1,FILE ='fm')
        REWIND (1)
        DO 50 J = 1, K+1
           FF = 0.5*(J-1)/K
           WRITE (1,130) FF, ABS(B(J))
50      CONTINUE
100     FORMAT ('LEAST SQUARE ERROR DESIGN OF A LOWPASS FILTER')
110     FORMAT ('ENTER: N, L, FP, DC, K')
120     FORMAT (9F8.5)
130     FORMAT (5X,F15.6,E15.8)
        GOTO 5
        END
C------------------END OF MAIN PROGRAM-------------------------
C
        SUBROUTINE FREQ(X,A,N,K)
        REAL X(1), A(1)
C
        Q  = 3.141592654/K
        AM = (N+1)*0.5
        M  = (N+1)/2
        N2 = N/2
        DO 20 J = 1, K+1
           AT = 0
           IF (AM.EQ.M) AT = 0.5*X(M)
           DO 10 I = 1, N2
              AT = AT + X(I)*COS(Q*(AM-I)*(J-1))
10         CONTINUE
           A(J) = 2*AT
20      CONTINUE
        RETURN
        END
```

3. A FORTRAN PROGRAM FOR LINEAR-PHASE LOW-PASS FIR FILTER DESIGN USING A LEAST SQUARED ERROR CRITERION AND A TRANSITION REGION

The FORTRAN program is a system for designing a length-N, linear-phase, low-pass FIR digital filter with a frequency response that is a LS error approximation to an ideal low-pass frequency response with a transition region as shown in Fig. 3.1b amd Fig. 3.16. A choice of a P-order spline or a raised cosine transition function is given. After the filter is designed, its frequency response is calculated for analysis. The basic theory, formulas, variable names, and references are chosen to follow the development in Sections 3.2.1 and 3.2.2.

The main program starts with a section that takes input specifications from the terminal. The length of the desired FIR filter is entered as N. Then follows FP, the pass-band edge in Hertz and FS, the stop-band edge in hertz; both assume a sampling rate of 1.0. Next, an integer TP is entered to specify the type of transition function to be used. Enter 0 for a raised cosine transition function, 1 for a first-order spline (straight-line) transition function, 2 for a second-order spline, and, in general, an integer P for a P-order spline transition function. A value for K is entered to set the number of frequencies at which the frequency response is evaluated in the analysis section of the program.

The subroutine LS( ) calculates the impulse response from (3.29) for an ideal rectangular low-pass response with a band edge at the average of FP and FS. The subroutine WGT( ) calculates the various weight functions by using (3.32) and (3.33), which result from the transition functions, and multiplies them by the ideal impulse response to give the actual filter coefficients. The first half of the symmetric impulse response is written to the terminal as the coefficients of the designed filter. The frequency response of the filter is calculated at K equally spaced frequencies by the subroutine FREQ( ), which implements (3.2) and (3.5); and these values are written to the file fm.

This program gives an optimal LS error approximation to the ideal low-pass filter with a transition region. It is fast and very accurate, even for large N, because the design is analytical rather than numerical. The transition region allows us to control the design by the specifications. The choice of transition function and the length-N control the residual approximation error.

```
C          LEAST SQUARE ERROR FIR FILTERS WITH A TRANSITION REGION
C          DESIGN PROGRAM FOR A LINEAR PHASE LOWPASS FILTER
C          FILTER LENGTH = N
C             PASSBAND EDGE IN HERTZ = FP
C             STOPBAND EDGE IN HERTZ = FS,   FOR SAMPLING RATE = 1
C          TRANSITION TYPE = TP: 0. RAISED COSINE. 1. LINEAR, 2. 2ND ORDER,
C                  3. 3RD ORDER, 4. 4TH ORDER, ETC.;   FP=FS=NO TRANSITION
C             FREQUENCY RESPONSE CALCULATED AT K POINTS
C          C.S. BURRUS,   RICE UNIVERSITY,   JAN 1987
C-------------------------------------------------------------
       INTEGER TP
       REAL X(101), B(1001)
```

```
C------------------INPUT SPECIFICATIONS-----------------
        WRITE (6,100)
    7   WRITE (6,110)
        READ (5,*) N, FP, FS, TP, K
        M  = (N+1)/2
        FQ = FS - FP
        FR = FS + FP
C-----------------DESIGN-----------------------------
        CALL LS (X, N, FR, FQ)
        CALL WGT (X, N, TP, FR, FQ)
C----------------OUTPUT------------------------------
        WRITE (6,120)(X(J),J=1,M)
        CALL FREQ(X,B,N,K)
        OPEN (1,FILE ='fm')
        REWIND(1)
        DO 50 J = 1, K+1
            F = 0.5*(J-1)/K
            WRITE (1,130) F, ABS(B(J))
   50   CONTINUE
  100   FORMAT ('LEAST SQUARE DESIGN WITH A TRANSITON REGION')
  110   FORMAT ('ENTER:  N, FP, FS, TP, K')
  120   FORMAT (9F8.5)
  130   FORMAT (5X,F15.8,E15.8)
        GOTO 7
        END
C--------------END OF MAIN PROGRAM--------------------
C
        SUBROUTINE LS(X,N,FP,FQ)
        REAL X(1)
C
        P  = 3.141592654
        M  = (N+1)/2
        AM = (N+1.0)/2.0
        N2 = N/2
        IF (M.EQ.AM) X(M) = FP
        DO 10 J = 1, N2
            Q = P*(J - AM)
            X(J) = (SIN(FP*Q))/Q
   10   CONTINUE
        RETURN
        END



C-------------------WEIGHTS----------------------
        SUBROUTINE WGT(X,N,TP,FP,FQ)
        INTEGER TP
        REAL X(1)
C
        AM = (N+1.0)/2.0
        P  = 3.141592654
        Q  = P*FQ
        IF (FQ.EQ.0) GOTO 7
        IF (TP.EQ.0) GOTO 5
C---------------TP ORDER SPLINE---------------
    1   DO 11 J = 1, AM-1
            Q1  = Q*(J-AM)/TP
            WT  = (SIN(Q1)/Q1)**TP
            X(J) = WT*X(J)
   11   CONTINUE
        RETURN
C-------------RAISED COSINE--------------------
    5   DO 15 J = 1, AM-1
            WT  = COS(Q*(J-AM))
            IF (ABS(WT).GT.1.0E-6) GOTO 13
            WT  = P/4.0
            GOTO 14
   13       WT  = WT/(1-(2*FQ*(J-AM))**2)
   14       X(J) = WT*X(J)
   15   CONTINUE
    7   RETURN
        END
```

```
C-------------FREQUENCY RESPONSE-------------
        SUBROUTINE FREQ(X,A,N,K)
        REAL X(1), A(1)
C
        Q   = 3.141592654/K
        AM  = (N+1)*0.5
        M   = (N+1)/2
        N2  = N/2
        DO 20 J = 1, K+1
          AT = 0
          IF (AM.EQ.M) AT = 0.5*X(M)
          DO 10 I = 1, N2
            AT = AT + X(I)*COS(Q*(AM-I)*(J-1))
10        CONTINUE
          A(J) = 2*AT
20      CONTINUE
        RETURN
        END
```

4. A FORTRAN PROGRAM FOR LINEAR-PHASE LOW-PASS FIR FILTER DESIGN USING A LEAST SQUARED ERROR CRITERION AND OPTIONAL WINDOWS

The FORTRAN program is a system for designing a length-N linear-phase, low-pass FIR digital filter with a frequency response that is an LS error approximation to an ideal low-pass frequency response. Filters designed with this error criterion exhibit a ripple or oscillation in their frequency response, called the Gibbs phenomenon. When this oscillation is undesirable, we can use one of five window functions to reduce it. After the filter is designed, its frequency response is calculated for analysis. The basic theory, formulas, variable names, and references are chosen to follow the development in Sections 3.2.1 and 3.2.3.4.

The main program starts with a section that takes input specifications from the terminal. The length of the desired FIR filter is entered as N. Then follows FP, the cutoff frequency or band edge in Hertz, where we assume a rate of 1.0. Next, an integer TP is entered to the type of window function used to truncate the ideal, infinitely long, impulse response. Enter 0 for a rectangular window (simple truncation), 1 for Bartlett, 2 for Hanning, 3 for Hamming, 4 for Blackman, or 5 for Kaiser. If the Kaiser window is chosen, a parameter BET with values in the range of 2 to 15 must be entered. A value for K is entered to set the number of frequencies at which the frequency response is evaluated in the analysis section of the program.

The subroutine LS( ) calculates the ideal impulse response from (3.29). The subroutine WIND( ) calculates the various window functions, using (3.40) through (3.42), and multiplies them by the ideal impulse response to give the actual filter coefficients. The Kaiser window requires calculation of a Bessel function, which is done by the FORTRAN function IO( ). The first half of the symmetric impulse response is written to the terminal as the coefficients of the designed filter. The frequency response of the filter is calculated at K equally spaced frequencies by the subroutine FREQ( ), which implements (3.2) and (3.5); these values are written to the file fm.

This program gives an optimal least mean squared error approximation to the ideal low-pass filter. It is fast and accurate, even for large N, because design is analytical rather than numerical. Unfortunately, therefore, it is not easy to modify it to approximate something other than a simple, constant-magnitude ideal. The windows give some flexibility, but they destroy the optimality. The Kaiser window is probably the most useful tradeoff of ripple and transition width. An input section could be added to take ripple and transition width and calculate N and BET.

```
C          LEAST SQUARE ERROR AND WINDOWS FOR FIR FILTERS
C          DESIGN PROGRAM FOR A LINEAR PHASE LOWPASS FILTER
C          FILTER LENGTH = N
C          BAND EDGE IN HERTZ = FP,    FOR SAMPLING RATE = 1
C          WINDOW TYPE:     TP = 0.RECTANGULAR, 1.BARTLETT,
C                        2.HANNING, 3.HAMMING, 4.BLACKMAN, 5.KAISER
C          KAISER WINDOW PARAMETER = BETA
C          FREQUENCY RESPONSE CALCULATED AT K POINTS
C          C.S. BURRUS,    JAN 1987
C-----------------------------------------------------------
           INTEGER TP
           REAL X(101), B(1001)
C------------------------INPUT SPECIFICATIONS-----------------
           WRITE (6,100)
    5      WRITE (6,110)
           READ (5,*) N, FP, TP, K
           M  = (N+1)/2
C--------------------DESIGN----------------------------------
           CALL LS (X,N,FP)
           CALL WIND (X, N, TP)
C--------------------OUTPUT----------------------------------
           WRITE (6,120)(X(J),J=1,M)
           CALL FREQ(X,B,N,K)
           OPEN (1,FILE ='fm')
           REWIND(1)
           DO 50 J = 1, K+1
               F = 0.5*(J-1)/K
               WRITE (1,130) F, ABS(B(J))
   50      CONTINUE
  100      FORMAT ('LEAST SQUARE DESIGN WITH WINDOWS')
  110      FORMAT ('ENTER:  N, FP, TP, K')
  120      FORMAT (9F8.5)
  130      FORMAT (5X,F15.8,E15.8)
           GOTO 5
           END
C---------------END OF MAIN PROGRAM--------------------
C
           SUBROUTINE LS(X,N,FP)
           REAL X(1)
C
           P  = 3.141592654
           M  = (N+1)/2
           AM = (N+1.0)/2.0
           N2 = N/2
           IF (M.EQ.AM) X(M) = 2.0*FP
           DO 10 J = 1, N2
               Q = P*(J - AM)
               X(J) = (SIN(FP*2*Q))/Q
   10      CONTINUE
           RETURN
           END
```

```
C----------------WINDOWS----------------------
      SUBROUTINE WIND(X,N,TP)
      INTEGER TP
      REAL X(1)
C
      IF (TP.EQ.0) RETURN
      AM = (N+1.0)/2.0
      Q  = 3.141592654/AM
      Q1 = 3.141592654/(AM-1)
      GOTO (1,2,3,4,5), TP
      RETURN
C----------------BARTLETT--------------------
  1   DO 11 J = 1, AM
          WT   = J/AM
          X(J) = WT*X(J)
 11   CONTINUE
      RETURN
C----------------HANNING---------------------
  2   DO 12 J = 1, AM
          WT   = 0.5 - 0.5*COS(J*Q)
          X(J) = WT*X(J)
 12   CONTINUE
      RETURN
C----------------HAMMING---------------------
  3   DO 13 J = 1, AM
          WT   = 0.54 - 0.46*COS((J-1)*Q1)
          X(J) = WT*X(J)
 13   CONTINUE
      RETURN
C----------------BLACKMAN--------------------
  4   DO 14 J = 1, AM
          WT   = 0.42 - 0.5*COS(J*Q) + 0.08*COS(2*J*Q)
          X(J) = WT*X(J)
 14   CONTINUE
      RETURN
C----------------KAISER----------------------
  5   PRINT *,'ENTER BETA'
      READ *,BET
      FIO1 = FIO(BET)
      DO 15 J = 1, AM
          ARG  = BET*SQRT(1-((AM-J)/(AM-1))**2)
          WT   = FIO(ARG)/FIO1
          X(J) = WT*X(J)
 15   CONTINUE
      RETURN
      END
```

```
C----------------IO FUNCTION-----------------
      FUNCTION FIO(Z)
      Y = Z/2.0
      E = 1.0
      D = 1.0
      DO 10 J = 1, 25
          D  = D*Y/J
          D2 = D*D
          E  = E + D2
          IF (D2.LT.E*1E-7) GOTO 15
 10   CONTINUE
      PRINT *,'IO FAILED TO CONVERGE'
 15   FIO = E
      RETURN
      END
```

```
C----------------------------------------------------
      SUBROUTINE FREQ(X,A,N,K)
      REAL X(1), A(1)
C
      Q  =  3.141592654/K
      AM = (N+1)*0.5
      M  = (N+1)/2
      N2 = N/2
      DO 20 J = 1, K+1
         AT = 0
         IF (AM.EQ.M) AT = 0.5*X(M)
         DO 10 I = 1, N2
            AT = AT + X(I)*COS(Q*(AM-I)*(J-1))
  10     CONTINUE
         A(J) = 2*AT
  20  CONTINUE
      RETURN
      END
```

5. A FORTRAN PROGRAM FOR LINEAR-PHASE FIR DIFFERENTIATOR DESIGN USING A LEAST SQUARED ERROR CRITERION

The FORTRAN program is a system for designing a length-N, linear-phase, FIR digital differentiator with a frequency response that is an LS error approximation to an ideal differentiator frequency response. Filters designed with this error criterion exhibit a ripple or oscillation in their frequency response, called the Gibb phenomenon, where a discontinuity exists. This phenomenon occurs for type 3 filters with an odd N but not for type 4 filters with an even N. After the filter is designed, its frequency response is calculated for analysis. The basic theory, formulas, variable names, and references are chosen to follow the development in Sections 3.2.1 and 3.2.2.

The main program starts with a section that takes input specifications from the terminal. The length of the desired FIR filter is entered as N. A value for K is entered to set the number of frequencies at which the frequency response is evaluated in the analysis section of the program.

The program determines if N is odd or even and branches to the appropriate section for design of a type 3 or type 4 filter. If N is odd, the ideal impulse response is calculated from (3.30); if it is even, (3.31) is used. The first half of the odd-symmetric impulse response is written to the terminal as the coefficients of the designed filter. The frequency response of the filter is calculated at K equally spaced frequencies by the subroutine FREQ( ), which implements (2.28) and (2.29) or (3.13); these values are written to the file fm.

This program gives an optimal least mean squared error approximation to the ideal differentiator. It is fast and accurate, even for large N, because the design is analytical rather than numerical. If a wide-band differentiator is desired, an even length, type 4 filter should be used. The Gibbs phenomenon for the type 3 differentiator is significant near $\omega = \pi$ because of the discontinuity that must occur there (see Figs. 2.4 and 3.15).

If a narrow-band differentiator is desired, a new design formula should be derived that combines a differentiator and a filter. If possible, a transition region should be included to reduce the Gibbs effect.

```
C           LEAST SQUARE ERROR FOR FIR DIFFERENTIATOR
C           DESIGN PROGRAM FOR A LINEAR PHASE DIFFERENTIATOR
C             FILTER LENGTH = N
C             TYPE = TP: 3. ODD N,    4. EVEN N
C             FREQUENCY RESPONSE CALCULATED AT K POINTS
C           C.S. BURRUS,    JAN 1987
C--------------------------------------------------------------
            REAL X(1001), B(1001)
C-------------------INPUT SPECIFICATIONS-----------------
            WRITE (6,100)
      5     WRITE (6,110)
            READ (5,*) N, K
            M   = (N+1)/2
            AM  = (N+1.0)/2.0
            N2  = N/2
            P   = 3.141592654
            IF (M.NE.AM) GOTO 11
C-------------TYPE 3, ODD N-----------------------
            DO 10 J = 1, M-1
               X(J) = (COS(P*(M-J)))/(M-J)
      10    CONTINUE
            X(M) = 0.0
            GOTO 12
C-------------TYPE 4, EVEN N----------------------
      11    DO 12 J = 1, N2
               Q = P*(J - AM)
               X(J) = (SIN(Q))/(Q*(J-AM))
      12    CONTINUE
C------------------OUTPUT--------------------------
            WRITE (6,120)(X(J),J=1,M)
            CALL FREQ(X,B,N,K)
            OPEN (1,FILE ='fm')
            REWIND(1)
            DO 50 J = 1, K+1
               F = 0.5*(J-1)/K
               WRITE (1,130) F, ABS(B(J))
      50    CONTINUE
     100    FORMAT ('LEAST SQUARE DESIGN OF A DIFFERENTIATOR')
     110    FORMAT ('ENTER:  N, K')
     120    FORMAT (9F8.5)
     130    FORMAT (5X,F15.8,E15.8)
            GOTO 5
            END
C-----------------------------------------------------
            SUBROUTINE FREQ(X,A,N,K)
            REAL X(1), A(1)
C
            Q   = 3.14159264/K
            AM  = (N+1)*0.5
            N2  = N/2
            DO 20 J = 1, K+1
               AT = 0
               DO 10 I = 1, N2
                  AT = AT + X(I)*SIN(Q*(AM-I)*(J-1))
      10       CONTINUE
               A(J) = 2*AT
      20    CONTINUE
            RETURN
            END
```

6. A FORTRAN PROGRAM FOR MULTIBAND LINEAR-PHASE FIR DESIGN USING THE CHEBYSHEV ERROR CRITERION AND THE PARKS–McCLELLAN ALGORITHM

A listing of a slightly modified version of the program EQFIR, which appeared in the IEEE Press book *Programs for Digital Signal Processing* is included here with the permission of the IEEE. It is called the Parks–McClellan algorithm. Some modifications have been made in the input and output parts of the program so that the input is read from the terminal and the output is stored in a file, PM.LST, and written to the screen.

This program implements the theory described in Section 3.3. The main program starts with a section that takes input from the terminal. The prompts that appear on the screen are listed and described.

Enter filter length, type, no. of bands

The filter length is read in as "nfilt". The three possible filter types (see the following discussion) are

Type 1   Multiple pass-band/stop-band filter
Type 2   Differentiator
Type 3   Hilbert transform filter

The number of bands is two for a low-pass filter since the transition band is not counted. For a bandpass filter with one pass band, the number of bands is three: a stop band, the pass band, and the second stop band.

Enter band edges

The band edges should be entered in fractions of the sampling frequency in ascending order. A low-pass filter has four bandedges: 0.0, the end of the pass band, the beginning of the stop band, and 0.5. Bandpass filters have correspondingly more band edges.

Enter desired value in each band

For a low-pass filter the desired value is 1.0 in the first band and 0.0 in the second band.

Enter weight in each band

When the weights are equal to 1.0 in all bands, the error will be the same for all bands. Since the program minimizes the *weighted* error, a larger weight gives a smaller error. For example, if the pass-band weight is 1.0 and the stop-band weight is 10.0, the stop-band error will be one tenth as large as the pass-band error.

The filter type in the program (JTYPE) has a different meaning than that used for the word "type" in Section 2.2.1. When JTYPE = 1, the parameter $m = 0$ in (3.43). If JTYPE = 1 and the filter length is odd, the program designs a type 1 filter described in Section 2.2.1. When JTYPE = 2 and the filter length is even, a type 2 filter is designed. If JTYPE = 3, the parameter $m = 1$ in (3.43), and the program designs a type 3 filter for odd length or a type 4 filter for even length. For JTYPE = 1 and JTYPE = 3, the desired function is a constant in the specified bands. The choice JTYPE = 2 was included for the design of differentiators where the desired function is linear with a specified slope. When JTYPE = 1 or JTYPE = 3, the weight is a constant in the specified bands. However, when JTYPE = 2, the weight is inversely proportional to the value of the desired function, to give a constant percent error.

After the filter for Example 3.9 was designed, the following information appeared on the screen:

```
                    finite impulse response (fir)
                    linear phase digital filter design
                       remes exchange algorithm
                          bandpass filter

                       filter length = 21
```

	band 1	band 2
lower band edge	0.	0.3700000
upper band edge	0.3300000	0.5000000
desired value	1.0000000	0.
weighting	1.0000000	1.0000000
deviation	0.0988697	0.0988697
deviation in db	0.8189238	−20.0987320

```
          *****filter specs are in the file pm.1st*****
          *****impulse response is in file r.dat*****
```

In applications, such as equalization, when the desired function is not constant or linear and the desired weight is not constant or proportional to the desired value, the user can write new functions EFF for the desired function and WATE for the special weighting function.

The examples in Section 3.3.3 were designed with this program. Each example illustrates a particular feature in the design of filters using this program. Before using this program, read the guidelines in Section 3.3.3.

It is always a good idea to calculate and plot the frequency response of a filter before using it in an application. The filter is optimum in terms of the mathematical criteria used in the theory of Chebyshev approximation, but it may have exactly the expected frequency response.

Usually the program will be run several times to get an appropriate filter. The formulas for filter length, given in Section 3.3.4, are only approximate starting points. Often the length must be changed and the program run again to meet specifications. The choice of weight function is often an iterative procedure as well. As described in Section 3.3.1, the maximum value of the product of the weight $W(f)$ and the error is minimized. A small weight allows larger errors, whereas a large weight allows smaller errors.

A convenient way to use the program is to make up an input file and, in UNIX, redirect the input. The input file for Example 3.9 was

21	1	2	
0.0	0.33	0.37	0.5
1.0	0.0		
1.0			

The resulting output file, pm.lst, is

• •

finite impulse response (fir)
linear phase digital filter design
remez exchange algorithm
bandpass filter

filter length = 21

*****impulse response*****

$$
\begin{aligned}
h(\ 1) &= \ \ 0.18255439e-01 = h(21) \\
h(\ 2) &= \ \ 0.55136755e-01 = h(20) \\
h(\ 3) &= -0.40910728e-01 = h(19) \\
h(\ 4) &= \ \ 0.14930855e-01 = h(18) \\
h(\ 5) &= \ \ 0.27568584e-01 = h(17) \\
h(\ 6) &= -0.59407797e-01 = h(16) \\
h(\ 7) &= \ \ 0.44841841e-01 = h(15) \\
h(\ 8) &= \ \ 0.31902660e-01 = h(14) \\
h(\ 9) &= -0.14972545e+00 = h(13) \\
h(10) &= \ \ 0.25687239e+00 = h(12) \\
h(11) &= \ \ 0.69994062e+00 = h(11)
\end{aligned}
$$

	band 1	band 2
lower band edge	0.	0.3700000
upper band edge	0.3300000	0.5000000
desired value	1.0000000	0.
weighting	1.0000000	1.0000000
deviation	0.0988697	0.0988697
deviation in db	0.8189238	−20.0987320

```
c
c------------------------------------------------------------------------
c main program: fir linear phase filter design program
c
c authors: james h. mcclellan
c          schlumber well services
c          12125 technology blvd.
c          austin, texas 78759
c
c          thomas w. parks
c          school of electrical engineering
c          cornell university
c          ithaca, new york 14853
c
c          lawrence r. rabiner
c          bell laboratories
c          murray hill, new jersey 07974
c
c modified for terminal input by t.w. parks
c
c input:
c   nfilt-- filter length
c   jtype-- type of filter
c            1 = multiple passband/stopband filter
c            2 = differentiator
c            3 = hilbert transform filter
c   nbands-- number of bands
c
c   edge(2*nbands)-- array; lower and upper edges for each band
c                    with a maximum of 10 bands.
c
c   fx(nbands)-- desired function array (or desired slope if a
c                differentiator) for each band.
c
c   wtx(nbands)-- weight function array in each band.  for a
c                 differentiator, the weight function is inversely
c                 proportional to f.
c
c   sample input data setup:
c       32,1,3
c       0.0,0.1,0.2,0.35,0.425,0.5
c       0.0,1.0,0.0
c       10.0,1.0,10.0
c     this data specifies a length 32 bandpass filter with
c     stopbands 0 to 0.1 and 0.425 to 0.5, and passband from
c     0.2 to 0.35 with weighting of 10 in the stopbands and 1
c     in the passband.
c
c     the following input data specifies a length 32 fullband
c     differentiator with slope 1 and weighting of 1/f.
c       32,2,1
c       0,0.5
c       1.0
c       1.0
c
c------------------------------------------------------------------------
c
        common pi2,ad,dev,x,y,grid,des,wt,alpha,iext,nfcns,ngrid
        common /oops/niter,iout
        dimension iext(66),ad(66),alpha(66),x(66),y(66)
        dimension h(66),hh(132)
        dimension des(1045),grid(1045),wt(1045)
        dimension edge(20),fx(10),wtx(10),deviat(10)
        double precision pi2,pi
        double precision ad,dev,x,y
        double precision gee,d
        integer bd1,bd2,bd3,bd4
        data bd1,bd2,bd3,bd4/1hb,1ha,1hn,1hd/
        input= 5
```

```
      iout= 3
      pi=4.0*datan(1.0d0)
      pi2=2.0d00*pi
c
c   the program is set up for a maximum length of 128, but
c   this upper limit can be changed by redimensioning the
c   arrays iext, ad, alpha, x, y, h to be nfmax/2 + 2.
c   the arrays des, grid, and wt must dimensioned
c   16(nfmax/2 + 2).
c
      nfmax=128
  100 continue
      open(3,file ='pm.lst')
      open(4,file ='r.dat')
      jtype=0
c
c   program input section
c
      write(*,104)
  104 format(3x,'Enter filter length, type, no. of bands')
      read *, nfilt,jtype,nbands
      if(nfilt.eq.0)stop
      if(nfilt.le.nfmax.or.nfilt.ge.3) go to 115
      call error
      stop
  115 if(nbands.le.0) nbands=1
c
c   grid density is assumed to be 16
c
      lgrid=16
      jb=2*nbands
      write(*,120)
  120 format(3x,'Enter band edges')
      read *, (edge(j),j=1,jb)
      write(*,121)
  121 format(3x,'Enter desired value in each band')
      read *, (fx(j),j=1,nbands)
      write(*,122)
  122 format(3x,'Enter weight in each band')
      read *, (wtx(j),j=1,nbands)
      if(jtype.gt.0.and.jtype.le.3) go to 125
      call error
      stop
  125 neg=1
      if(jtype.eq.1) neg=0
      nodd=nfilt/2
      nodd=nfilt-2*nodd
      nfcns=nfilt/2
      if(nodd.eq.1.and.neg.eq.0) nfcns=nfcns+1
c
c   set up the dense grid.  the number of points in the grid
c   is (filter length + 1)*grid density/2
c
      grid(1)=edge(1)
      delf=lgrid*nfcns
      delf=0.5/delf
      if(neg.eq.0) go to 135
      if(edge(1).lt.delf) grid(1)=delf
  135 continue
      j=1
      l=1
      lband=1
  140 fup=edge(l+1)
  145 temp=grid(j)
c
c   calculate the desired magnitude response and the weight
c   function on the grid
c
      des(j)=eff(temp,fx,wtx,lband,jtype)
```

292

```
      wt(j)=wate(temp,fx,wtx,lband,jtype)
      j=j+1
      grid(j)=temp+delf
      if(grid(j).gt.fup) go to 150
      go to 145
  150 grid(j-1)=fup
      des(j-1)=eff(fup,fx,wtx,lband,jtype)
      wt(j-1)=wate(fup,fx,wtx,lband,jtype)
      lband=lband+1
      l=l+2
      if(lband.gt.nbands) go to 160
      grid(j)=edge(l)
      go to 140
  160 ngrid=j-1
      if(neg.ne.nodd) go to 165
      if(grid(ngrid).gt.(0.5-delf)) ngrid=ngrid-1
  165 continue
c
c  set up a new approximation problem which is equivalent
c  to the original problem
c
      if(neg) 170,170,180
  170 if(nodd.eq.1) go to 200
      do 175 j=1,ngrid
      change=dcos(pi*grid(j))
      des(j)=des(j)/change
  175 wt(j)=wt(j)*change
      go to 200
  180 if(nodd.eq.1) go to 190
      do 185 j=1,ngrid
      change=dsin(pi*grid(j))
      des(j)=des(j)/change
  185 wt(j)=wt(j)*change
      go to 200
  190 do 195 j=1,ngrid
      change=dsin(pi2*grid(j))
      des(j)=des(j)/change
  195 wt(j)=wt(j)*change
c
c  initial guess for the extremal frequencies--equally
c  spaced along the grid
c
  200 temp=float(ngrid-1)/float(nfcns)
      do 210 j=1,nfcns
      xt=j-1
  210 iext(j)=xt*temp+1.0
      iext(nfcns+1)=ngrid
      nm1=nfcns-1
      nz=nfcns+1
c
c  call the remes exchange algorithm to do the approximation
c  problem
c
      call remes
c
c  calculate the impulse response.
c
      if(neg) 300,300,320
  300 if(nodd.eq.0) go to 310
      do 305 j=1,nm1
      nzmj=nz-j
  305 h(j)=0.5*alpha(nzmj)
      h(nfcns)=alpha(1)
      go to 350
  310 h(1)=0.25*alpha(nfcns)
      do 315 j=2,nm1
      nzmj=nz-j
      nf2j=nfcns+2-j
  315 h(j)=0.25*(alpha(nzmj)+alpha(nf2j))
```

```
          h(nfcns)=0.5*alpha(1)+0.25*alpha(2)
          go to 350
  320 if(nodd.eq.0) go to 330
          h(1)=0.25*alpha(nfcns)
          h(2)=0.25*alpha(nm1)
          do 325 j=3,nm1
          nzmj=nz-j
          nf3j=nfcns+3-j
  325 h(j)=0.25*(alpha(nzmj)-alpha(nf3j))
          h(nfcns)=0.5*alpha(1)-0.25*alpha(3)
          h(nz)=0.0
          go to 350
  330 h(1)=0.25*alpha(nfcns)
          do 335 j=2,nm1
          nzmj=nz-j
          nf2j=nfcns+2-j
  335 h(j)=0.25*(alpha(nzmj)-alpha(nf2j))
          h(nfcns)=0.5*alpha(1)-0.25*alpha(2)
c
c   program output section.
c
c since iout=3, the output is written to file 'pm.lst'
  350 write(iout,360)
  360 format(1h1, 70(1h*)/15x,29hfinite impulse response (fir)/
     113x,34hlinear phase digital filter design/
     217x,24hremes exchange algorithm)
          if(jtype.eq.1) write(iout,365)
  365 format(22x,15hbandpass filter/)
          if(jtype.eq.2) write(iout,370)
  370 format(22x,14hdifferentiator/)
          if(jtype.eq.3) write(iout,375)
  375 format(20x,19hhilbert transformer/)
          write(iout,378) nfilt
  378 format(20x,16hfilter length = ,i3/)
c for screen output, the impulse response is not written
          if (iout.eq.6) go to 457
          write(iout,380)
  380 format(15x,28h***** impulse response *****)
          do 381 j=1,nfcns
          k=nfilt+1-j
          if(neg.eq.0) write(iout,382) j,h(j),k
          if(neg.eq.1) write(iout,383) j,h(j),k
  381 continue
  382 format(13x,2hh(,i2,4h) = ,e15.8,5h = h(,i3,1h))
  383 format(13x,2hh(,i2,4h) = ,e15.8,6h = -h(,i3,1h))
          if(neg.eq.1.and.nodd.eq.1) write(iout,384) nz
  384 format(13x,2hh(,i2,8h) =   0.0)
c
c now to write impulse response to file 'r.dat'
c write the first half of the response
          do 785 i=1,nfcns
          hh(i)=h(i)
  785 continue
c
          if(neg.eq.0.and.nodd.eq.0)then
c here neg=0 and nodd=0 for bandpass even length filter
c nfcns=nfilt/2
              do 786 n=1,nfcns
              hh(nfcns+n)=h(nfcns-n+1)
  786         continue
              do 787 i=1,nfcns*2
              write(4,*) hh(i)
  787         continue
c
          else if(neg.eq.0.and.nodd.eq.1) then
c here neg=0 and nodd=1 for bandpass odd length filter
c nfcns=nfilt/2 + 1
              do 788 n=1,nfcns-1
              hh(nfcns+n)=h(nfcns-n)
```

```
      788         continue
                  do 789 i=1,nfcns*2-1
                  write(4,*) hh(i)
      789         continue
c
              else if(neg.eq.1.and.nodd.eq.0) then
c neg=1 for diff and hilbert, nodd=0 for even length
                  do 800 n=1,nfcns
                  hh(nfcns+n)=-h(nfcns-n+1)
      800         continue
                  do 801 i=1,nfcns*2
                  write(4,*) hh(i)
      801         continue
c
              else if(neg.eq.1.and.nodd.eq.1) then
c neg=1 for diff and hilbert, nodd=1 for odd length
              h(nfcns+1)=0
                  do 802 n=1,nfcns
                  hh(nfcns+n+1)=-h(nfcns-n+1)
      802         continue
                  do 803 i=1,nfcns*2+1
                  write(4,*) hh(i)
      803         continue
          end if
      457 do 450 k=1,nbands,4
          kup=k+3
          if(kup.gt.nbands) kup=nbands
          write(iout,385) (bd1,bd2,bd3,bd4,j,j=k,kup)
      385 format(24x,4(4a1,i3,7x))
          write(iout,390) (edge(2*j-1),j=k,kup)
      390 format(2x,15hlower band edge,5f14.7)
          write(iout,395) (edge(2*j),j=k,kup)
      395 format(2x,15hupper band edge,5f14.7)
          if(jtype.ne.2) write(iout,400) (fx(j),j=k,kup)
      400 format(2x,13hdesired value,2x,5f14.7)
          if(jtype.eq.2) write(iout,405) (fx(j),j=k,kup)
      405 format(2x,13hdesired slope,2x,5f14.7)
          write(iout,410) (wtx(j),j=k,kup)
      410 format(2x,9hweighting,6x,5f14.7)
          do 420 j=k,kup
      420 deviat(j)=dev/wtx(j)
          write(iout,425) (deviat(j),j=k,kup)
      425 format(2x,9hdeviation,6x,5f14.7)
          if(jtype.ne.1) go to 450
          do 430 j=k,kup
      430 deviat(j)=20.0*alog10(deviat(j)+fx(j))
          write(iout,435) (deviat(j),j=k,kup)
      435 format(2x,15hdeviation in db,5f14.7)
      450 continue
          do 452 j=1,nz
          ix=iext(j)
      452 grid(j)=grid(ix)
c extremal frequencies not written out in this version
c         write(iout,455) (grid(j),j=1,nz)
      455 format(/2x,47hextremal frequencies--maxima of the error curve/
     1 (2x,5f12.7))
c         write(iout,460)
      460 format(1x,70(1h*)/)
          if (iout.eq.6) go to 461
          iout=6
          go to 350
      461 write(*,460)
          write(*,462)
      462 format(10x,45h*****filter specs are in the file pm.lst*****/)
      458 write(*,459)
      459 format(10x,45h******impulse response is in file r.dat******/)
          stop
          end
c
```

```
c-----------------------------------------------------------------------
c function: eff
c    function to calculate the desired magnitude response
c    as a function of frequency.
c    an arbitrary function of frequency can be
c    approximated if the user replaces this function
c    with the appropriate code to evaluate the ideal
c    magnitude.  note that the parameter freq is the
c    value of normalized frequency needed for evaluation.
c-----------------------------------------------------------------------
c
      function eff(freq,fx,wtx,lband,jtype)
      dimension fx(5),wtx(5)
      if(jtype.eq.2) go to 1
      eff=fx(lband)
      return
    1 eff=fx(lband)*freq
      return
      end
c
c-----------------------------------------------------------------------
c function: wate
c    function to calculate the weight function as a function
c    of frequency.  similar to the function eff, this function can
c    be replaced by a user-written routine to calculate any
c    desired weighting function.
c-----------------------------------------------------------------------
c
      function wate(freq,fx,wtx,lband,jtype)
      dimension fx(5),wtx(5)
      if(jtype.eq.2) go to 1
      wate=wtx(lband)
      return
    1 if(fx(lband).lt.0.0001) go to 2
      wate=wtx(lband)/freq
      return
    2 wate=wtx(lband)
      return
      end
c
c-----------------------------------------------------------------------
c subroutine: error
c    this routine writes an error message if an
c    error has been detected in the input data.
c-----------------------------------------------------------------------
c
      subroutine error
      common /oops/niter,iout
      write(iout,1)
    1 format(44h *********** error in input data **********)
      return
      end
c
c-----------------------------------------------------------------------
c subroutine: remes
c    this subroutine implements the remes exchange algorithm
c    for the weighted chebyshev approximation of a continuous
c    function with a sum of cosines.  inputs to the subroutine
c    are a dense grid which replaces the frequency axis, the
c    desired function on this grid, the weight function on the
c    grid, the number of cosines, and an initial guess of the
c    extremal frequencies.  the program minimizes the chebyshev
c    error by determining the best location of the extremal
c    frequencies (points of maximum error) and then calculates
c    the coefficients of the best approximation.
c-----------------------------------------------------------------------
c
      subroutine remes
      common pi2,ad,dev,x,y,grid,des,wt,alpha,iext,nfcns,ngrid
```

```
      common /oops/niter,iout
      dimension iext(66),ad(66),alpha(66),x(66),y(66)
      dimension des(1045),grid(1045),wt(1045)
      dimension a(66),p(65),q(65)
      double precision pi2,dnum,dden,dtemp,a,p,q
      double precision dk,dak
      double precision ad,dev,x,y
      double precision gee,d
c
c  the program allows a maximum number of iterations of 25
c
      itrmax=25
      devl=-1.0
      nz=nfcns+1
      nzz=nfcns+2
      niter=0
  100 continue
      iext(nzz)=ngrid+1
      niter=niter+1
      if(niter.gt.itrmax) go to 400
      do 110 j=1,nz
      jxt=iext(j)
      dtemp=grid(jxt)
      dtemp=dcos(dtemp*pi2)
  110 x(j)=dtemp
      jet=(nfcns-1)/15+1
      do 120 j=1,nz
  120 ad(j)=d(j,nz,jet)
      dnum=0.0
      dden=0.0
      k=1
      do 130 j=1,nz
      l=iext(j)
      dtemp=ad(j)*des(l)
      dnum=dnum+dtemp
      dtemp=float(k)*ad(j)/wt(l)
      dden=dden+dtemp
  130 k=-k
      dev=dnum/dden
c        write(iout,131) dev
c intermeditate deviations not written in this version
  131 format(1x,12hdeviation = ,f12.9)
      nu=1
      if(dev.gt.0.0) nu=-1
      dev=-float(nu)*dev
      k=nu
      do 140 j=1,nz
      l=iext(j)
      dtemp=float(k)*dev/wt(l)
      y(j)=des(l)+dtemp
  140 k=-k
      if(dev.gt.devl) go to 150
      call ouch
      go to 400
  150 devl=dev
      jchnge=0
      k1=iext(1)
      knz=iext(nz)
      klow=0
      nut=-nu
      j=1
c
c  search for the extremal frequencies of the best
c  approximation
c
  200 if(j.eq.nzz) ynz=comp
      if(j.ge.nzz) go to 300
      kup=iext(j+1)
      l=iext(j)+1
```

297

```
       nut=-nut
       if(j.eq.2) y1=comp
       comp=dev
       if(l.ge.kup) go to 220
       err=gee(l,nz)
       err=(err-des(l))*wt(l)
       dtemp=float(nut)*err-comp
       if(dtemp.le.0.0) go to 220
       comp=float(nut)*err
210    l=l+1
       if(l.ge.kup) go to 215
       err=gee(l,nz)
       err=(err-des(l))*wt(l)
       dtemp=float(nut)*err-comp
       if(dtemp.le.0.0) go to 215
       comp=float(nut)*err
       go to 210
215    iext(j)=l-1
       j=j+1
       klow=l-1
       jchnge=jchnge+1
       go to 200
220    l=l-1
225    l=l-1
       if(l.le.klow) go to 250
       err=gee(l,nz)
       err=(err-des(l))*wt(l)
       dtemp=float(nut)*err-comp
       if(dtemp.gt.0.0) go to 230
       if(jchnge.le.0) go to 225
       go to 260
230    comp=float(nut)*err
235    l=l-1
       if(l.le.klow) go to 240
       err=gee(l,nz)
       err=(err-des(l))*wt(l)
       dtemp=float(nut)*err-comp
       if(dtemp.le.0.0) go to 240
       comp=float(nut)*err
       go to 235
240    klow=iext(j)
       iext(j)=l+1
       j=j+1
       jchnge=jchnge+1
       go to 200
250    l=iext(j)+1
       if(jchnge.gt.0) go to 215
255    l=l+1
       if(l.ge.kup) go to 260
       err=gee(l,nz)
       err=(err-des(l))*wt(l)
       dtemp=float(nut)*err-comp
       if(dtemp.le.0.0) go to 255
       comp=float(nut)*err
       go to 210
260    klow=iext(j)
       j=j+1
       go to 200
300    if(j.gt.nzz) go to 320
       if(k1.gt.iext(1)) k1=iext(1)
       if(knz.lt.iext(nz)) knz=iext(nz)
       nut1=nut
       nut=-nu
       l=0
       kup=k1
       comp=ynz*(1.00001)
       luck=1
310    l=l+1
       if(l.ge.kup) go to 315
```

```
          err=gee(l,nz)
          err=(err-des(l))*wt(l)
          dtemp=float(nut)*err-comp
          if(dtemp.le.0.0) go to 310
          comp=float(nut)*err
          j=nzz
          go to 210
  315   luck=6
          go to 325
  320   if(luck.gt.9) go to 350
          if(comp.gt.y1) y1=comp
          k1=iext(nzz)
  325   l=ngrid+1
          klow=knz
          nut=-nut1
          comp=y1*(1.00001)
  330   l=l-1
          if(l.le.klow) go to 340
          err=gee(l,nz)
          err=(err-des(l))*wt(l)
          dtemp=float(nut)*err-comp
          if(dtemp.le.0.0) go to 330
          j=nzz
          comp=float(nut)*err
          luck=luck+10
          go to 235
  340   if(luck.eq.6) go to 370
          do 345 j=1,nfcns
          nzzmj=nzz-j
          nzmj=nz-j
  345   iext(nzzmj)=iext(nzmj)
          iext(1)=k1
          go to 100
  350   kn=iext(nzz)
          do 360 j=1,nfcns
  360   iext(j)=iext(j+1)
          iext(nz)=kn
          go to 100
  370   if(jchnge.gt.0) go to 100
c
c   calculation of the coefficients of the best approximation
c   using the inverse discrete fourier transform
c
  400   continue
          nm1=nfcns-1
          fsh=1.0e-06
          gtemp=grid(1)
          x(nzz)=-2.0
          cn=2*nfcns-1
          delf=1.0/cn
          l=1
          kkk=0
          if(grid(1).lt.0.01.and.grid(ngrid).gt.0.49) kkk=1
          if(nfcns.le.3) kkk=1
          if(kkk.eq.1) go to 405
          dtemp=dcos(pi2*grid(1))
          dnum=dcos(pi2*grid(ngrid))
          aa=2.0/(dtemp-dnum)
          bb=-(dtemp+dnum)/(dtemp-dnum)
  405   continue
          do 430 j=1,nfcns
          ft=j-1
          ft=ft*delf
          xt=dcos(pi2*ft)
          if(kkk.eq.1) go to 410
          xt=(xt-bb)/aa
          xt1=sqrt(1.0-xt*xt)
          ft=atan2(xt1,xt)/pi2
  410   xe=x(l)
```

```
          if(xt.gt.xe) go to 420
          if((xe-xt).lt.fsh) go to 415
          l=l+1
          go to 410
   415    a(j)=y(l)
          go to 425
   420    if((xt-xe).lt.fsh) go to 415
          grid(1)=ft
          a(j)=gee(1,nz)
   425    continue
          if(l.gt.1) l=l-1
   430    continue
          grid(1)=gtemp
          dden=pi2/cn
          do 510 j=1,nfcns
          dtemp=0.0
          dnum=j-1
          dnum=dnum*dden
          if(nm1.lt.1) go to 505
          do 500 k=1,nm1
          dak=a(k+1)
          dk=k
   500    dtemp=dtemp+dak*dcos(dnum*dk)
   505    dtemp=2.0*dtemp+a(1)
   510    alpha(j)=dtemp
          do 550 j=2,nfcns
   550    alpha(j)=2.0*alpha(j)/cn
          alpha(1)=alpha(1)/cn
          if(kkk.eq.1) go to 545
          p(1)=2.0*alpha(nfcns)*bb+alpha(nm1)
          p(2)=2.0*aa*alpha(nfcns)
          q(1)=alpha(nfcns-2)-alpha(nfcns)
          do 540 j=2,nm1
          if(j.lt.nm1) go to 515
          aa=0.5*aa
          bb=0.5*bb
   515    continue
          p(j+1)=0.0
          do 520 k=1,j
          a(k)=p(k)
   520    p(k)=2.0*bb*a(k)
          p(2)=p(2)+a(1)*2.0*aa
          jm1=j-1
          do 525 k=1,jm1
   525    p(k)=p(k)+q(k)+aa*a(k+1)
          jp1=j+1
          do 530 k=3,jp1
   530    p(k)=p(k)+aa*a(k-1)
          if(j.eq.nm1) go to 540
          do 535 k=1,j
   535    q(k)=-a(k)
          nf1j=nfcns-1-j
          q(1)=q(1)+alpha(nf1j)
   540    continue
          do 543 j=1,nfcns
   543    alpha(j)=p(j)
   545    continue
          if(nfcns.gt.3) return
          alpha(nfcns+1)=0.0
          alpha(nfcns+2)=0.0
          return
          end
c
c------------------------------------------------------------------------
c function: d
c     function to calculate the lagrange interpolation
c     coefficients for use in the function gee.
c------------------------------------------------------------------------
c
```

```
      double precision function d(k,n,m)
      common pi2,ad,dev,x,y,grid,des,wt,alpha,iext,nfcns,ngrid
      dimension iext(66),ad(66),alpha(66),x(66),y(66)
      dimension des(1045),grid(1045),wt(1045)
      double precision ad,dev,x,y
      double precision q
      double precision pi2
      d=1.0
      q=x(k)
      do 3 l=1,m
      do 2 j=1,n,m
      if(j-k)1,2,1
    1 d=2.0*d*(q-x(j))
    2 continue
    3 continue
      d=1.0/d
      return
      end
c
c-----------------------------------------------------------------------
c function: gee
c    function to evaluate the frequency response using the
c    lagrange interpolation formula in the barycentric form
c
c-----------------------------------------------------------------------
c
      double precision function gee(k,n)
      common pi2,ad,dev,x,y,grid,des,wt,alpha,iext,nfcns,ngrid
      dimension iext(66),ad(66),alpha(66),x(66),y(66)
      dimension des(1045),grid(1045),wt(1045)
      double precision p,c,d,xf
      double precision pi2
      double precision ad,dev,x,y
      p=0.0
      xf=grid(k)
      xf=dcos(pi2*xf)
      d=0.0
      do 1 j=1,n
      c=xf-x(j)
      c=ad(j)/c
      d=d+c
    1 p=p+c*y(j)
      gee=p/d
      return
      end
c
c-----------------------------------------------------------------------
c subroutine: ouch
c    writes an error message when the algorithm fails to
c    converge.  there seem to be two conditions under which
c    the algorithm fails to converge: (1) the initial
c    guess for the extremal frequencies is so poor that
c    the exchange iteration cannot get started, or
c    (2) near the termination of a correct design,
c    the deviation decreases due to rounding errors
c    and the program stops.  in this latter case the
c    filter design is probably acceptable, but should
c    be checked by computing a frequency response.
c
c-----------------------------------------------------------------------
c
      subroutine ouch
      common /oops/niter,iout
      write(iout,1)niter
    1 format(44h ********** failure to converge **********/
     141h0probable cause is machine rounding error/
     223h0number of iterations =,i4/
     339h0if the number of iterations exceeds 3,/
     450h0the design may be correct, but should be verified)
      return
      end
```

301

7-8 FORTRAN PROGRAMS FOR CHEBYSHEV AND LEAST SQUARED COMPLEX APPROXIMATION, SPECIALIZED FOR BANDPASS, DIFFERENTIATION, AND HILBERT TRANSFORMATION FILTERS WITH REDUCED DELAY

Program 7 is intended to minimize the maximum magnitude of the complex frequency response errors on a fine grid of frequencys for a length-N FIR filter with real coefficients. In the past bands, the desired frequency response has a constant magnitude for band pass and hilbert filters and a linear magnitude for differentiators. The desired phase is linear with a slope determined from the desired delay which is entered from the keyboard.

In Program 8 the sum of the squared magnitudes of complex frequency-response errors on a fine grid of frequencies is minimized in this program for a length-N FIR filter with real coefficients. In the pass band the desired function has a constant magnitude and a linear phase with a slope determined by the value of the desired delay, which is entered from the keyboard. Frequency bands can be separated by transition transition regions, and different weights can be used in various bands. The filter is not restricted to have exactly linear phase. Rather, the desired group delay is specified so that filters may be designed with less delay than the linear-phase filters.

The input for these programs is patterned after Program 6, the Parks–McClellan program. The input and output for program 8 are listed below with corresponding explanations: (They are similar for Program 7.)

Enter filter length, type, no. of bands

These entries are the same as for Program 6. The possible filter types are

Type 1   Multiple pass-band/stop-band filter
Type 2   Differentiator
Type 3   Hilbert transform filter

Enter desired group delay

This parameter determines the slope of the desired phase function. The delay is in normalized units. A desired delay of M corresponds to a delay of M samples.

Enter band edges (normalized in Hz).

The band edges are entered in ascending order as in Program 6.

Enter desired value of function in each band.

The usual values are 1.0 in the pass band and 0.0 in the stop band.

Enter weights for each band.

Just as in Program 6, the *weighted* error is minimized. A larger weight gives a smaller error.

To control group delay error, enter 1, else 0.

When a 1 is entered, the group delay error is reduced according to the weight entered in response to the next request.

Enter weight for group delay.

A larger weight gives a smaller group delay.

To control phase error, enter 1, else 0.

Since the group delay is determined by the phase *slope*, this additional weight is necessary if the phase error is important.

After the parameters are entered, the following outputs appear on the screen for a reduced delay version of example 3.9:

```
design in progress
        info =    0
finite impulse response (fir) ****************************************************
                              digital filter design
                          least square approximation
                                bandpass filter

                            filter length = 21

                      band 1              band 2              band
lower band edge       0.                  0.370000000
upper band edge       0.330000000         0.500000000
desired value         1.000000000         0.
weighting             1.000000000         1.000000000
----------------------------------------------------------------------
desired group delay   8.000000000
weight of group delay 1.000000000
****************************************************************************
                  *****filter specs are in the file l.lst*****
                  **filter impulse response is in file r.dat**
```

Program 8 uses the approach described in Section 3.2. First, a set of linear equations representing the desired filter characteristics is derived. These equations are then "solved" in the LS by the standard LINPACK QR decomposition subroutines DQRDC and DQRSL.

The details of the design are written out to the file named l.lst, and the impulse response alone is written to the file r.dat. Both the magnitude and delay (or phase) of the resulting response should be checked to see if they meet specifications, since this program minimizes the sum of the squared magnitudes of the complex frequency response errors.

The input parameters shown were used to design a length-21 filter with the same band edges as the linear-phase filter in Example 3.9, but with a desired delay of 8 samples, rather than the 10-sample delay resulting with exactly linear phase.

21	1	2	
8			
0.0	0.33	0.37	0.5
1.0	0.0		
1.0	1.0		
1			
1			
0			

The output of the design program, found in the file l.lst is

••

finite impulse response (fir)
digital filter design
least square approximation
bandpass filter

filter length = 21

*****impulse response*****
h(1) = −0.26436972e−01
h(2) = 0.11478931e−01
h(3) = 0.25567593e−01
h(4) = −0.57302951e−01
h(5) = 0.45461593e−01
h(6) = 0.29440641e−01
h(7) = −0.14849445e+00
h(8) = 0.25806873e+00
h(9) = 0.69688821e+00
h(10) = 0.25809250e+00
h(11) = −0.14683524e+00
h(12) = 0.27705259e−01
h(13) = 0.45791512e−01
h(14) = −0.55593180e−01
h(15) = 0.22229100e−01
h(16) = 0.15402893e−01

$$h(17) = -0.29811903e-01$$
$$h(18) = 0.25426015e-02$$
$$h(19) = -0.80981371e-03$$
$$h(20) = -0.22156683e-03$$
$$h(21) = 0.46910686e-03$$

	band 1	band 2
lower band edge	0.	0.370000000
upper band edge	0.330000000	0.500000000
desired value	1.000000000	0.
weighting	1.000000000	1.000000000

desired group delay	8,000000000
weight of group delay	1.000000000

Several modifications of these programs can be made. The input specification section of the program can be bypassed, and an arbitrary complex-valued desired frequency response can be used. This modification would be useful for designing equalizer filters.

Program 7

This program is intended to minimize the maximum magnitude of the complex frequency response errors on a fine grid of frequencies for a length-N FIR filter with real coefficients. In the passbands, the desired frequency response has a constant magnitude for bandpass and Hilbert filters and a linear magnitude for differentiators. The desired phase is linear with a slope determined from the desired delay which is entered from the keyboard.

```
c PROGRAM FOR THE DESIGN OF FIR FILTERS IN THE COMPLEX DOMAIN
c
c authors:   X. Chen
c            Department of Engineering
c            University of Denver
c            Denver, CO   80208
c
c            T. W. Parks
c            School of Electrical Engineering
c            Cornell University
c            Ithaca, NY   14853
c
c            See the paper "Design of FIR filters in the complex domain" by
c            X. Chen and T. W. Parks, to appear in IEEE-ASSP.
c
c            solve  min W(f)| f(z) - h(0) - h(1)*(1/z) - h(2)*(1/z**2) - ....|
c            where f(z(i))= u(z(i))+j v(z(i))    for   i=1,2,.......,m
c            and all the h(k) are real
c
c            normal equation is AH=B
c            A is stored as A'
c                m    discrete points of frequencies f
c                p    discrete points of auxiliary variable theta
c                     t(j)=(j-1)/2p
c                     theta(j)=pi2*t(j)           j=1,...,p
```

```
c                n       length of the filter
c         A=  ( a(s,k) ) is m*p+1 by n+3
c              a(s,k) = W(f(i)) cos(pi2 (k-1)f(i) - pi2 t(j))
c                  where      s=i+(j-1)m
c                             i=1,....,m
c                             j=1,....,p
c                             k=1,...,n
c              a(s,k) = 0      for k=n+1,n+2,n+3
c                              and for s=mp+1
c         H= transpose of ( h(1), h(2),  .... , h(n))
c         B= transpose of (b(1),.......b(m),0)
c                b(s)=W(fi)*(u(i)*cos(theta(j)) - v(i)*sin(theta(j)))
c
c         Uses Algorithm 495 by I. Barrodale and C. Phillips
c         in ACM Trans. math. Software, v.1, pp.264-270,1975.
c
c         Additional columns may be added to control error(s) of phase
c         or/and group delay.
c
          implicit real*8(a-h,o-z)
          double precision a(67,8193),b(8193),h(67)
          double precision grid(1024),des(1024),wt(1024),u(1024),v(1024)
          double precision edge(20),fx(10),wtx(10),wgd,deviat(10)
          character*6 ffile
          character*1 iout,igd,iph
c
          pi=3.141592653589793d0
          pi2=pi*2.d0
c         -----------------------------------------
c         program input section
c         (64*16=1024,   64*16*8=8192)
c
          nfmax=64


100       continue
          ifile=6
          write (6,90)
90        format('If printer output desired on this terminal',/,
     &    'enter y else n. (If n entered, prompts will still  ',/,
     &    'appear on this terminal)')
          read (5,96) iout
96        format(a1)
          if (iout.eq.'n') then
                  ifile=8
                  write (6,92)
92                format('Enter the desired file name (6 characters only).',/)
                  read (5,921) ffile
921               format(a6)
                  open (8,file=ffile)
                  rewind 8
                  write (6,94) ffile
94                format('Screen output now in  file ',a6,/)
                  end if
          write(6,104)
104       format('Type filter length, type,  bands, grid density')
          read(5,105) nfilt,jtype,nbands,lgrid
105       format(8i10)
          if(nfilt.gt.nfmax.or.nfilt.lt.1) call error
          if(nbands.le.0) nbands=1
          write(6,106)
```

```
106        format('Type auxiliary grid density')
           read(5,105) lp
c
c          main grid density is assumed to be 8 unless specified
c          otherwise
c
           if(lgrid.le.0) lgrid=8
c          auxilary grid density is assumed to be 8 unless specified
c          otherwise
c
           if(lp.le.0) lp=8
           write(6,111)
111        format('Type desired group delay')
           read(5,110) slop
           slop=slop*pi2
112        jb=2*nbands
           write(6,109)
109        format('Type band edges.')
           read(5,110) (edge(j),j=1,jb)
110        format(8f10.0)
           write(6,107)
107        format('Type desired value of function in each band.')
           read(5,110) (fx(j),j=1,nbands)
           write(6,108)
108        format('Type weights for each band.')
           read(5,110) (wtx(j),j=1,nbands)
           write(6,115)
115        format('If error in group delay is desired to control',/,
&          'enter y else n.')
           read (5,113) igd
113        format(a1)
           if (igd.eq.'n') go to 122
119        write(6,114)
114        format('Type weight for group delay.')


           read(5,110) wgd
122        write(6,123)
123        format('If error in phase is desired to control',/,
&          'enter y else n.')
           read(5,113) iph
           if(iph.eq.'n') go to 120
           write(6,117)
117        format('Type weight for phase error')
           read(5,110) wph
120        if(jtype.eq.0) call error
c          -----------------------------------------------
c          set up the dense grid.  the number of points in the grid
c          is (filter length + 1)*grid density
c
           grid(1)=edge(1)
           delf=lgrid*nfilt
           delf=0.5/delf
           if(edge(1).lt.delf) grid(1)=delf
135        continue
           j=1
           l=1
           lband=1
140        fup=edge(l+1)
145        temp=grid(j)
c          -----------------------------------------------
c          calculate the desired frequency response and the weight
c          function on the grid
```

```
c
          call eff(temp,slop,fx,wtx,lband,jtype,des(j),u(j),v(j))
          wt(j)=wate(temp,fx,wtx,lband,jtype)
          j=j+1
          grid(j)=temp+delf
          if(grid(j).gt.fup) go to 150
          go to 145
150       klk=j-1
          grid(klk)=fup
          call eff(fup,slop,fx,wtx,lband,jtype,des(klk),u(klk),v(klk))
          wt(j-1)=wate(fup,fx,wtx,lband,jtype)
          lband=lband+1
          l=l+2
          if(lband.gt.nbands) go to 160
          grid(j)=edge(l)
          go to 140
160       ngrid=j-1
c         if(grid(ngrid).gt.(0.5-delf)) ngrid=ngrid-1
c         ------------------------------------------
c         derive linear equations for the complex approximation
c
          do 300 j=1,lp
             theta=pi*(j-1)/dfloat(lp)
             tc=dcos(theta)
             ts=dsin(theta)
             do 280 i=1,ngrid
                mt=i+(j-1)*ngrid
                b(mt) = u(i)*tc-v(i)*ts
                b(mt) = wt(i)*b(mt)
                do 260 k=1,nfilt
                   temp=pi2*(k-1)*grid(i)-theta
                   a(k,mt)=wt(i)*dcos(temp)
260             continue




280          continue
300       continue
c
          ms=ngrid*up
c
          if(igd.eq.'n') go to 320
c         ------------------------------------------
c         adding constraints on the errors in group delay
c
          delay=slop/pi2
301       do 310 i=1,ngrid
             if(des(i).lt.1.e-5) go to 310
             ms=ms+1
             b(ms) = 0.d0
             theta=-slop*grid(i)
             if(jtype.eq.2) theta=theta+pi/2.d0
             if(jtype.eq.3) theta=theta-pi/2.d0
             do 305 k=1,nfilt
                temp=pi2*(k-1)*grid(i)+theta
                a(k,ms)=wgd*(k-1.d0-delay)*dcos(temp)
305          continue
310       continue
c
320       if(iph.eq.'n') go to 340
```

308

```
c          ------------------------------------------------
c          adding constraints on the error in phase
c
           theta=-pi/2.d0
           do 330 i=1,ngrid
              if(des(i).lt.1.e-5) go to 330
              ms=ms+1
              b(ms) = 0.d0
              do 325 k=1,nfilt
                 temp=pi2*(k-1)*grid(i)+theta
                 a(k,ms)=wph*dsin(temp)
325           continue
330        continue
c          ------------------------------------------------
c
340        if(ms.gt.8192) call error
           ndim=67
           mdim=8193
           tol=1.d-15
           relerr=0.d0
c
           call cheb(ms,nfilt,mdim,ndim,a,b,tol,relerr,h,nrank,resmax,iter,ncode)
c          ------------------------------------------------
c          program output section.
c
350        write(ifile,360)
360        format(70(1h*)//25x,'finite impulse response (fir)'/
     &     20x,'quasi-linear phase digital filter design'/
     &     25x,'complex approximation'/
     &     30x,'algorithm 495'/)
           if(jtype.eq.1) write(ifile,365)
365        format(30x,'bandpass filter'/)
           if(jtype.eq.2) write(ifile,370)
370        format(30x,'differentiator'/)
           if(jtype.eq.3) write(ifile,375)
375        format(30x,'hilbert transformer'/)




           write(ifile,378) nfilt
378        format(30x,'filter length = ',i3/)
           write(ifile,380)
380        format(22x,'***** impulse response *****')
           do 381 j=1,nfilt
           write(ifile,382) j,h(j)
381        continue
382        format(20x,'h(',i3,') = ',e15.8)
           do 450 k=1,nbands,4
           kup=k+3
           if(kup.gt.nbands) kup=nbands
           write(ifile,385) (j,j=k,kup)
385        format(/24x,4('band',i3,8x))
           write(ifile,390) (edge(2*j-1),j=k,kup)
390        format(2x,'lower band edge',5f15.9)
           write(ifile,395) (edge(2*j),j=k,kup)
395        format(2x,'upper band edge',5f15.9)
           if(jtype.ne.2) write(ifile,400) (fx(j),j=k,kup)
400        format(2x,'desired value',2x,5f15.9)
           if(jtype.eq.2) write(ifile,405) (fx(j),j=k,kup)
405        format(2x,'desired magn. slope',2x,5f15.9)
           write(ifile,410) (wtx(j),j=k,kup)
```

309

```
410       format(2x,'weighting',6x,5f15.9)
          do 412 j=k,kup
              deviat(j)=resmax/wtx(j)
              if(fx(j).gt.0.001 .and. jtype.eq.2) deviat(j)=deviat(j)/2.
412       continue
          write(ifile,414) (deviat(j),j=k,kup)
414       format(2x,'deviation',6x,5f15.9)
          do 416 j=k,kup
416           deviat(j)=20.d0*dlog10(deviat(j)+fx(j))
          write(ifile,418) (deviat(j),j=k,kup)
418       format(2x,'deviation in dB',5f15.9)
          write(ifile,419)
          if(igd.eq.'n') go to 470
          write(ifile,408) wgd
408       format(2x,'weight of group delay',f15.9)
          devgd=resmax/wgd
          write(ifile,406) devgd
406       format(2x,'error of group delay',1x,f15.9)
470       if(iph.eq.'n') go to 480
          write(ifile,471) wph
471       format(2x,'weight of phase error',f15.9)
          devph=resmax/wph
          write(ifile,472) devph
472       format(2x,'error of phase',6x,f15.9)
480       slop=slop/pi2
          write(ifile,407) slop
407       format(2x,'desired group delay',2x,f15.9)
          write(ifile,419)
419       format(2x,66(1h-))
          write(ifile,420) lgrid
420       format(2x,'grid density',2x,i7)
          write(ifile,422) lp
422       format(2x,'auxiliary grid',2x,i6)
          write(ifile,425) iter
425       format(2x,'iterations',6x,i5)
          write(ifile,430) ncode
430       format(2x,'exit code',7x,i5)
450       continue



900       write(ifile,460)
460       format(/,70(1h*)/)
          close (8)
          stop
          end
c         ************************************************************
c
          subroutine eff(temp,slop,fx,wtx,lband,jtype,des,u,v)
c
c         function to calculate the desired frequency response
c         as a function of frequency.
c
          implicit double precision(a-h,o-z)
          dimension fx(10),wtx(10)
          if (jtype.eq.2) go to 2
          if (jtype.eq.3) go to 3
          t=-temp*slop
          des=fx(lband)
          u=des*dcos(t)
          v=des*dsin(t)
          return
```

```fortran
2         t=temp*slop
          des=fx(lband)*temp*2.d0*3.141592653589793d0
          u=des*dsin(t)
          v=des*dcos(t)
          return
3         des=fx(lband)
          if (des.lt.1.d-5) go to 4
          t=-temp*slop
          u=des*dsin(t)
          v=-des*dcos(t)
          return
4         u=0.d0
          v=0.d0
          return
          end
c      -------------------------------------------------------
c

          function wate(temp,fx,wtx,lband,jtype)
c
c         function to calculate the weight function as a function
c         of frequency.
c
          implicit double precision(a-h,o-z)
          dimension fx(5),wtx(5)
          if(jtype.eq.2) go to 1
          wate=wtx(lband)
          return
1         if(fx(lband).lt.0.0001) go to 2
          wate=wtx(lband)/temp
          return
2         wate=wtx(lband)
          return
          end
c      ------------------------------------------------
          subroutine error
          write(ifile,1)
1         format(' *********** error in input data **********')
          stop .
          end
```

Program 8

```
c main program: fir least-square design program
c
c authors: x. chen and t.w. parks
c             department of electrical and computer engineering
c             rice university
c             houston, texas 77251
c
c input:
c         nfilt-- filter length
c
c         jtype-- type of filter
c                     1 - multiple passband/stopband filter
c                     2 - differentiator
c                     3 - Hilbert transform filter
c
c         nbands-- number of bands
c
c         lgrid-- grid density set to 8 unless otherwise specified
c
c         edge(2*nbands)-- bandedge array, lower and upper edges for
c                          each band with a maximum of 10 bands.
c
c         fx(nbands)-- desired function array for each band.
c
c         wtx(nbands)-- weight function array in each band.
c
c         slope-- desired group delay
c
c         wgd--    weight for group delay error
c
c         wph--    weight for phase error
c
c   sample data setup:
c         32,1,3
c         15.5
c         0.0,0.1,0.2,0.35,0.425,0.5
c         0.0,1.0,0.0
c         10.0,1.0,10.0
c         y
c         10.0
c         n
c   this data specifies a length 32 bandpass filter with
c   desired group delay of 15.5,
c   stopbands 0 to 0.1 and 0.425 to 0.5, and passband from
c   0.2 to 0.35 with a weight on the error magnitude-squared of 10
c   in the stopbands and 1 in the passband. the group delay is
c   weighted by 10.0 and the phase error is not directly controlled.
c
        implicit real*8(a-h,o-z)
        double precision a(4200,64),b(4200),h(64)
        double precision work(4200),qy(4200),qty(4200),ah(4200)
        double precision qraux(64),eor(4200)
        double precision grid(1025),des(1025),wt(1025),u(1025),v(1025)
        double precision edge(20),fx(10),wtx(10)
        integer jpvt(64)
c
        pi=3.141592653589793d0
        pi2=pi*2.d0
c
        open(3, file='l.lst')
        iout=3
c
c       program input section
c
        nfmax=64
100     continue
```

312

```
         write(*,104)
104      format(3x,'Enter filter length, type, no. of bands')
         read(*,*) nfilt,jtype,nbands
         if(nfilt.gt.nfmax.or.nfilt.lt.3) call error
c
c        main grid density is assumed to be 16 unless specified
c        otherwise
c
         lgrid=16
c
         write(*,111)
111      format(3x,'Enter desired group delay')
         read(*,*) slope
         slope=slope*pi2
112      jb=2*nbands
         write(*,109)
109      format(3x'Enter band edges (normalized in Hz.).')
         read(*,*) (edge(j),j=1,jb)
         write(*,107)
107      format(3x,'Enter desired value of function in each band.')
         read(*,*) (fx(j),j=1,nbands)
         write(*,108)
108      format(3x,'Enter weights for each band.')
         read(*,*) (wtx(j),j=1,nbands)
         write(*,115)
115      format(3x,'To control group delay error enter 1, else 0.')
         read (*,*) igd
         if (igd.eq.0) go to 122
119      write(*,114)
114      format(3x,'Enter weight for group delay.')
         read(*,*) wgd
122      write(*,123)
123      format(3x,'To control phase error enter 1, else 0.')
         read(*,*) iph
         if(iph.eq.0) go to 120
         write(*,117)
117      format(3x,'Enter weight for phase error')
         read(*,*) wph
120      if(jtype.eq.0) call error
         write(*,121)
121      format(3x,'design in progress')
c
c        set up the dense grid.  the number of points in the grid
c        is filter length * grid density (lgrid)
c
         grid(1)=edge(1)
         delf=lgrid*nfilt
         delf=0.5/delf
         if(edge(1).lt.delf) grid(1)=delf
135      continue
         j=1
         l=1
         lband=1
140      fup=edge(l+1)
145      temp=grid(j)
c
c
c        calculate the desired frequency response and the weight
c        function on the grid
c
         call eff(temp,slope,fx,wtx,lband,jtype,des(j),u(j),v(j))
         wt(j)=wate(temp,fx,wtx,lband,jtype)
         j=j+1
         grid(j)=temp+delf
         if(grid(j).gt.fup) go to 150
         go to 145
```

```
    150       klk=j-1
              grid(klk)=fup
              call eff(fup,slope,fx,wtx,lband,jtype,des(klk),u(klk),v(klk))
              wt(j-1)=wate(fup,fx,wtx,lband,jtype)
              lband=lband+1
              l=l+2
              if(lband.gt.nbands) go to 160
              grid(j)=edge(l)
              go to 140
    160       ngrid=j-1
              if(grid(ngrid).gt.(0.5-delf)) ngrid=ngrid-1
c
c             derive linear equations
c
              do 300 j=1,2
                 theta=pi*(j-1)/2.d0
                 tc=dcos(theta)
                 ts=dsin(theta)
                 do 280 i=1,ngrid
                    mt=i+(j-1)*ngrid
                    b(mt) = u(i)*tc-v(i)*ts
                    b(mt) = wt(i)*b(mt)
                    do 260 k=1,nfilt
                       temp=pi2*(k-1)*grid(i)-theta
                       a(mt,k)=wt(i)*dcos(temp)
    260             continue
    280          continue
    300       continue
c
              delay=slope/pi2
c
              ms=ngrid*2
c
              if(igd.eq.0) go to 320
c
c             adding constraint on the error of group delay
c
    301       do 310 i=1,ngrid
                 if(des(i).lt.1.e-5) go to 310
                 ms=ms+1
                 b(ms) = 0.d0
                 theta=-slope*grid(i)
                 if(jtype.eq.2) theta=theta+pi/2.d0
                 if(jtype.eq.3) theta=theta-pi/2.d0
                 do 305 k=1,nfilt
                    temp=pi2*(k-1)*grid(i)+theta
                    a(ms,k)=wgd*(k-1.d0-delay)*dcos(temp)
    305          continue
    310       continue
c
    320       if(iph.eq.0) go to 340
c
c
c             adding phase error constraint
c
              theta=-pi/2.d0
              do 330 i=1,ngrid
                 if(des(i).lt.1.e-5) go to 330
                 ms=ms+1
                 b(ms) = 0.d0
                 do 325 k=1,nfilt
                    temp=pi2*(k-1)*grid(i)+theta
                    a(ms,k)=wph*dsin(temp)
    325          continue
    330       continue
c
```

```
c
340     if(ms.gt.4699) call error

        call dqrdc(a,4200,ms,nfilt,qraux,jpvt,work,0)
        info=1
        call dqrsl(a,4200,ms,nfilt,qraux,b,qy,qty,h,eor,ah,110,info)
c
        open(9, file='error')
        rewind(9)
        write(9,342) (eor(i),i=1,ms)
342     format(10x,e15.6)
        close(9)
        write(*,343) info
343     format(10x,'info=',i5)
        if(info.ne.0) stop
c
c
c       program output section.
c
c       first the impulse response is written to file 'r.dat'
c
        open(9, file='r.dat')
        rewind(9)
        do 345 i=1,nfilt
        write(9,*) h(i)
345     continue
        close(9)
c
c
c       the output is written to file 'l.lst' then to the screen
c
350     write(iout,360)
360     format(1x,70(1h*)/22x,29hfinite impulse response (fir)/
     125x,21hdigital filter design/
     223x,26hleast square approximation)
        if(jtype.eq.1) write(iout,365)
365     format(28x,'bandpass filter'/)
        if(jtype.eq.2) write(iout,370)
370     format(28x,'differentiator'/)
        if(jtype.eq.3) write(iout,375)
375     format(28x,'hilbert transformer'/)
        write(iout,378) nfilt
378     format(25x,'filter length = ',i3/)
c for screen output, the impulse response is not written
        if(iout.eq.6) go to 457
        write(iout,380)
380     format(20x,'***** impulse response *****')
        do 381 j=1,nfilt
        write(iout,382) j,h(j)
381     continue
382     format(20x,'h(',i3,') = ',e15.8)
457     do 450 k=1,nbands,4
        kup=k+3
        if(kup.gt.nbands) kup=nbands
        write(iout,385) (j,j=k,kup)
385     format(/24x,4('band',i3,8x))
        write(iout,390) (edge(2*j-1),j=k,kup)
390     format(2x,'lower band edge',5f15.9)
        write(iout,395) (edge(2*j),j=k,kup)
395     format(2x,'upper band edge',5f15.9)
        if(jtype.ne.2) write(iout,400) (fx(j),j=k,kup)
400     format(2x,'desired value',2x,5f15.9)
        if(jtype.eq.2) write(iout,405) (fx(j),j=k,kup)
405     format(2x,'magn.  slope ',2x,5f15.9)
        write(iout,410) (wtx(j),j=k,kup)
410     format(2x,'weighting',6x,5f15.9)
```

```
            write(iout,419)
            write(iout,407) delay
407         format(2x,'desired group delay',2x,f15.9)
            if(igd.eq.0) go to 470
            write(iout,408) wgd
408         format(2x,'weight of group delay',f15.9)
470         if(iph.eq.0) go to 450
            write(iout,471) wph
471         format(2x,'weight of phase error',f15.9)
450         continue
            write(iout, 460)
419         format(2x,66(1h-))
460         format(/,70(1h*)/)
            if (iout.eq.6) go to 461
            iout=6
            go to 350
461         write(*,462)
462         format(10x,44h*****filter specs are in the file l.lst*****/)
            write(*,463)
463         format(10x,44h**filter impulse response is in file r.dat**/)

            stop
            end
c
c
c           ********************************************************
c
            subroutine eff(temp,slope,fx,wtx,lband,jtype,des,u,v)
c
c           function to calculate the desired frequency response
c           as a function of frequency.
c
            implicit double precision(a-h,o-z)
            dimension fx(10),wtx(10)
            if (jtype.eq.2) go to 2
            if (jtype.eq.3) go to 3
            t=-temp*slope
            des=fx(lband)
            u=des*dcos(t)
            v=des*dsin(t)
            return
2           t=temp*slope
            des=fx(lband)*temp*2.d0*3.141592653589793d0
            u=des*dsin(t)
            v=des*dcos(t)
            return
3           des=fx(lband)
            if (des.lt.1.d-5) go to 4
            t=-temp*slope
            u=des*dsin(t)
            v=-des*dcos(t)
            return
4           u=0.d0
            v=0.d0
            return
            end
c           -------------------------------------------------------
c
            function wate(temp,fx,wtx,lband,jtype)
c
c           function to calculate the weight function as a function
c           of frequency.
c
            implicit double precision(a-h,o-z)
            dimension fx(5),wtx(5)
            if(jtype.eq.2) go to 1
```

316

```
        wate=wtx(lband)
        return
1       if(fx(lband).lt.0.0001) go to 2
        wate=wtx(lband)/temp
        return
2       wate=wtx(lband)
        return
        end
c       -----------------------------------------------
        subroutine error
        write(*,1)
1       format(' *********** error in input data **********')
        stop
        end
```

9. A FORTRAN PROGRAM FOR IIR FILTER DESIGN USING BUTTERWORTH- CHEBYSHEV- AND ELLIPTIC FUNCTION APPROXIMATIONS

The FORTRAN program is for designing digital and analog low-pass filters based on the classical Butterworth, Chebyshev I and II, and elliptic function approximations. These methods are magnitude approximations and produce minimum-phase filters. The basic theory, formulas, variable names, and references are chosen to follow the development in Sections 7.2 and 7.3.

The main program starts with a section that takes input specifications from the terminal. subroutine DFR( ) calculates the frequency response of the designed filter at a specified number of equally spaced points and writes them to a file named fm. The number of frequencies to be evaluated is first entered. Next, a choice is made between the four basic approximations: Butterworth, Chebyshev I or II, and elliptic function. An input is requested to determine whether a low-pass, high-pass, bandpass, or band-rejection filter is desired. Then a choice between analog and digital filter design is made. If the choice is digital, the bilinear transformation is used, and the sampling rate must be entered. If an elliptic filter was not chosen, the order is entered next. The pass-band and/or stop-band edges are entered in hertz; the maximum allowed pass-band ripple and/or the minimum allowed stop-band attenuation are entered in positive dB. Remember that for IIR filters, the pass-band ripple is defined as the total difference between the maximum and the minimum frequency responses over the pass band. This definition is in contrast to the FIR case, where the ripple is the difference between the maximum (or minimum) and the ideal responses.

The band edges are converted from Hertz to radians. Then if a digital filter is chosen, the frequencies are prewarped according to (7.112) by the PREWARP( ) function to the appropriate analog prototype values. If a low-pass frequency response was not chosen, the band edges WP and/or WS of the prototype low pass are calculated from the entered specifications by (7.97), (7.98), and (7.100).

The pole and zero locations for the Butterworth and Chebyshev I and II are calculated in the subroutine ROOTS1. Because the filter coefficients are real, the poles and zeros occur in complex conjugate pairs. The program calculates only one of the complex pair. It uses (7.11) for the Butterworth, (7.31) for the Chebyshev, and (7.47) and (7.48) for the Chebyshev II in the DO 15 loop. For the Chebyshev I and II the parameter ε from (7.19) is calculated from (7.36) and then used to calculate v_0 by (7.29). The Butterworth roots are the foundation of this subroutine. If a Chebyshev filter is desired, these roots are modified by the hyperbolic functions of v_0. If a Chebyshev II filter is desired, these roots are "inverted" and the zero locations are calculated in the indented section at label 11. Note that the root locations are scaled to the proper location by the WP or WS band-edge frequencies. Also note the special case that must be considered for an odd order, which always has a single real root.

The pole and zero locations for the elliptic function filter are calculated in the subroutine ROOTS2. The parameter ε in (7.58) is calculated by (7.90). The order N is next determined from (7.93), which requires the calculation of four complete elliptic integrals. The modulus k is calculated from (7.92) and scaled to give the desired pass-band edge by WP. The complementary modulus k' is calculated from k by (7.92). The second modulus k_1, defined in (7.68), is calculated from (7.89) and its complement k_1' from (7.91). The complete elliptic integrals are calculated by the FORTRAN function CEI( ), which uses the arithmetic-geometric mean (AGM) implemented by the very efficient algorithm in procedure cel1, page 86 of Bulrisch.[22] The order is calculated from (7.93). At this point the approach in method A of Section 7.2.8 is taken, which calculates a new k_1 to satisfy (7.71). This calculation is done in the function FK( ) by using a ratio of the power series expansions 16.33.7 and 16.38.5 on page 579 of reference 21. This approach is also taken in reference 23 and gives the maximum attenuation in the stop band. At this point the program could easily be modified to take the approach of paragraph B or C to minimize the pass-band ripple or the transition bandwidth.

From the various parameters calculated from the input specifications, v_0 is computed from (7.82) and requires an inverse elliptic tangent function (elliptic integral of the first kind). This calculation is performed by the AGM procedure el1 on page 85 of reference 22 in the FORTRAN function ARCSC( ). The elliptic sine, cosine, and dn functions are next calculated by the subroutine ELP( ) which uses the sncndn procedure on page 89 of reference 22. In the DO 15 loop the zero locations are calculated from (7.79), and the pole locations from (7.84). These are scaled by WP to give the proper pass-band and stop-band band edges.

After the root locations for the prototype low-pass analog filter are calculated, the frequency transformations of (7.99) or (7.101) are made in subroutine FREQXFM( ) if the desired filter is a high-pass, bandpass, or band-rejection filter. The bandpass and band-rejection transformations double the order of the prototype filter.

Next, the root locations for the prewarped prototype analog filter are transformed to the digital filter root locations by the bilinear transformation described in Section 7.3.2. It is done by the subroutine BLT( ), which uses

(7.106). The zero and pole locations are displayed on the terminal by the subroutine PRNT( ). The root locations are converted into second-order cascade section parameters by the subroutine CASCAD( ). If the order is odd, there is one first order section. The rule for ordering the sections for the low-pass filter has the first section with the pole(s) and zero(s) farthest from the unit circle and nearest the real axis in the z plane and progresses to the last section with the poles nearest the unit circle and the zeros nearest those poles. That seems to give reasonable quantization and scaling performance. The ordering and pairing for the bandpass and band-rejection cases must be worked out separately. The frequency response of the filter is calculated from these cascade parameters in subroutine DFR( ), and the response is written to the file "fm" by the subroutine VIEW( ).

This program is intended to provide the basis for a flexible optimal filter design system as well as to illustrate the implementation of the theory in Sections 7.2 and 7.3. The numerical algorithms, mainly from reference 22, are the most accurate and efficient known to the authors. No approximations other than those necessary in the elliptic function algorithms are used. The structure of the program gives the user considerable control over the design of a filter, and it can be modified to fully implement the optimal design properties of the theory.

The input and output are primitive and may need to be customized by the user. If it is desired to specify gains and band edges rather than the order for the Butterworth and both Chebyshev filters, the equations for calculating order in (7.15), (7.38), and (7.54) can be added to the input section. For more control over elliptic function filter design, options can be added to allow a choice of methods A, B, or C of Section 7.2.8.

Some applications require the impulse-invariant method of converting the analog prototype into a digital one. The techniques of Section 7.3.1 could be added as an option to the bilinear transformation. If the parallel structures of Section 8.1.4 are preferred, a subroutine for calculating the residues of the poles could be added to evaluate the parallel structure coefficients. If analog filter design is desired, everything is calculated in the program; only an output section is needed. For some exacting applications, double-precision arithmetic must be incorporated, testing for convergence in the elliptic function algorithms made more stringent, and array sizes enlarged. The programs are also written in a style that allows easy conversion to other languages.

```
C        THIS IS A IIR FILTER DESIGN PROGRAM
C        FOR BUTTERWORTH, CHEBYSHEV, CHEBYSHEV II, & ELLIPTIC
C        FOR LOWPASS, HIGHPASS, BANDPASS, AND BANDREJECT RESPONSES
C        ANALOG AND DIGITAL FILTERS USING THE BLT
C            PASS AND STOPBAND EDGES ARE IN HERTZ FOR A SAMPLING
C            RATE OF 1. MAXIMUM PASSBAND RIPPLE AND MINIMUM
C            STOPBAND ATTENUATION ARE IN POSITIVE DB.
C        C. S. BURRUS,     RICE UNIVERSITY,    JAN 1987
C------------------------------------------------------------------
         DIMENSION PR(20), PI(20), ZR(20), ZI(20)
         DIMENSION B1(20),B2(20),A1(20),A2(20)
         DIMENSION FM(530)
         COMMON /PARM/R1,R2,WP,WS,N2,N,SR,KA,KAD,KOD,KF
         COMMON /ROOT/PR,PI,ZR,ZI
```

```
C---------INPUT SPECIFICATIONS, PREWARPING, AND PREFREQXFRMING-------
        PRINT *,'ENTER NUMBER OF FREQS TO DISPLAY'
        READ *,KK
10      PRINT *,'ENTER 1 FOR BW, 2 FOR CHEBY ,3 FOR ICHEBY, 4 FOR ELL'
        READ *,KA
        PRINT *,'ENTER 1 FOR LOWPASS, 2 FOR HP, 3 FOR BP, OR 4 FOR BR'
        READ *,KF
        PRINT *,'ENTER 1 FOR ANALOG, 2 FOR DIGITAL'
        READ *,KAD
        TP = 6.283185307179586
        IF (KAD.EQ.1) GOTO 12
        PRINT *,'ENTER SAMPLE RATE'
        READ *,SR
12      IF (KA.EQ.4) GOTO 20
        PRINT *,'ENTER THE ORDER'
        READ *,N
        IF (KF.GE.3) GOTO 15
        PRINT *,'ENTER THE BAND EDGE IN UN-NORMALIZED HZ'
        READ *,FP
        WP = PREWRP(TP*FP)
        IF (KF.EQ.2) WP = 1.0/WP
        IF (KA.EQ.1) GOTO 30
        PRINT *,'ENTER PASSBAND RIPPLE OR STOPBAND ATT IN POSITIVE DB'
        READ *,R1
        IF (KA.EQ.3) WS = WP
        GOTO 30
15      PRINT *,'ENTER THE LOWER & UPPER BAND EDGES IN HERTZ'
        READ *,F1,F2
        W1 = PREWRP(TP*F1)
        W2 = PREWRP(TP*F2)
        W0 = SQRT(W1*W2)
        WP = (W2*W2-W0*W0)/W2
        IF (KF.EQ.4) WP = 1/WP
        IF (KA.EQ.1) GOTO 30
         PRINT *,'ENTER PASSBAND RIPPLE OR STOPBAND ATT IN + DB'
        READ *,R1
        IF (KA.EQ.3) WS = WP
        GOTO 30
20      IF (KF.GE.3) GOTO 25



        PRINT *,'ENTER PASS AND STOPBAND EDGES IN UN-NORMALIZED HZ'
        READ *,FP,FS
        WP = PREWRP(TP*FP)
        WS = PREWRP(TP*FS)
        IF (KF.EQ.2) WP = 1.0/WP
        IF (KF.EQ.2) WS = 1.0/WS
        PRINT *,'ENTER PASSBAND RIPPLE AND STOPBAND ATTENUATION IN +DB'
        READ *,R1,R2
        GOTO 35
25      PRINT *,'ENTER F1,F2,F3,F4 FOR BP OR BR FREQS'
        READ *,F1,F2,F3,F4
        W1 = PREWRP(TP*F1)
        W2 = PREWRP(TP*F2)
        W3 = PREWRP(TP*F3)
        W4 = PREWRP(TP*F4)
        W0 = SQRT(W3*W2)
        WP = (W3*W3-W0*W0)/W3
        WS = (W4*W4-W0*W0)/W4
        WST= (W0*W0-W1*W1)/W1
        IF (WST.LT.WS) WS = WST
        IF (KF.EQ.3) GOTO 26
            W0 = SQRT(W1*W4)
            WP = W1/(W0*W0-W1*W1)
            WS = W2/(W0*W0-W2*W2)
            WST= W3/(W3*W3-W0*W0)
            IF (WST.LT.WS) WS = WST
```

320

```
26      PRINT *,'ENTER PASSBAND RIPPLE AND STOPBAND ATTENUATION IN + DB'
        READ *,R1,R2
        GOTO 35
C-------------BUTTERWORTH, CHEBYSHEV, AND ELLIPTIC FILTERS--
30      CALL ROOTS1
        GO TO 37
35      CALL ROOTS2
C---------HIGHPASS, BANDPASS, AND BAND REJECT XFORMS--------
37      IF (KF.EQ.1) GOTO 65
        CALL FREQXFM(W0,PR,PI)
        CALL FREQXFM(W0,ZR,ZI)
        IF (KF.EQ.2) GOTO 65
        N2 = N
        N  = 2*N
        KOD = 0
65      IF (KAD.EQ.1) GOTO 80
C-------------DIGITAL BILINEAR XFORM-----------------------
        CALL BLT(N2,SR,PR,PI)
        CALL BLT(N2,SR,ZR,ZI)
        PRINT *,'Z PLANE'
        CALL PRNT(N2,PR,PI,ZR,ZI)
C-------------CASCADE STRUCTURE AND FREQUENCY RESPONSE------
80      CALL CASCAD(PR,PI,ZR,ZI,B1,B2,A1,A2,G)
        IF (KAD.EQ.2) CALL DFR(KK,B1,B2,A1,A2,FM,G)
        IF (KAD.EQ.1) CALL AFR
        CALL VIEW(KK,FM,DB)
        GOTO 10
        END
C-------------END OF MAIN PROGRAM-------------------------




C--------------BW, CHEBY I&II POLE & ZERO LOCATIONS--------
        SUBROUTINE ROOTS1
        DIMENSION PR(20),PI(20),ZR(20),ZI(20)
        COMMON /PARM/R1,R2,WP,WS,N2,N,SR,KA,KAD,KOD,KF
        COMMON /ROOT/PR,PI,ZR,ZI
        ARCSNH(X) = ALOG(X+SQRT(X*X+1))
C
        E = SQRT(10.0**(0.1*R1) - 1)
        IF (KA.EQ.3) E = 1/E
        L  = 0
        N2 = (N+1)/2
        KOD = 1
        IF (MOD(N,2).EQ.0) KOD = 0
        IF (KOD.EQ.0) L = 1
        SM = 1.0
        CM = 1.0
        IF (KA.EQ.1) GO TO 10
        V0 = ARCSNH(1/E)/N
        SM = SINH(V0)
        CM = COSH(V0)
10      DO 15 J = 1, N2
        ARG = 1.570796326794897*L/N
        TR = -SM*COS(ARG)
        TI =  CM*SIN(ARG)
        ZR(J) = 0.0
        ZI(J) = 1E25
        IF (KA.EQ.3) GOTO 11
        PR(J) = WP*TR
        PI(J) = WP*TI
        GO TO 12
11      IF (L.NE.0) ZI(J) = WS/SIN(ARG)
        PR(J) = WS*TR/(TR*TR + TI*TI)
        PI(J) = WS*TI/(TR*TR + TI*TI)
12      L = L + 2
15      CONTINUE
        RETURN
        END
```

```
C----------ELLIPTIC FILTER POLE & ZERO LOCATIONS------
      SUBROUTINE ROOTS2
      REAL K, KC, KK, KKC, K1, K1C, KK1, KK1C
      DIMENSION PR(20),PI(20),ZR(20),ZI(20)
      COMMON /ROOT/PR,PI,ZR,ZI
      COMMON /PARM/R1,R2,WP,WS,N2,N,SR,KA,KAD,KOD,KF
      COMMON /ELP1/SN,CN,DN
C
      E = SQRT(10.0**(0.1*R1)-1)
         K    = WP/WS
         KC   = SQRT(1-K*K)
         K1   = E/SQRT(10.**(0.1*R2)-1)
         K1C  = SQRT(1-K1*K1)
         KK   = CEI(KC)
         KKC  = CEI(K)
         KK1  = CEI(K1C)
         KK1C = CEI(K1)
         XN   = KK*KK1C/KK1/KKC
         N    = INT(XN + 1.0)
            PRINT *,'N= ',N
         K1   = FK(N*KKC/KK)
         K1C  = SQRT(1-K1*K1)
         KK1  = CEI(K1C)
      L  = 0
      N2 = (N+1)/2
      KOD = 1
         IF(MOD(N,2).EQ.0) KOD = 0
         IF(KOD.EQ.0) L = 1
      V0 = (KK/KK1/N)*ARCSC(1/E,K1)
      CALL ELP(V0,K)
      SM = SN
      CM = CN
      DM = DN
      ZI(1) = 1E25
      DO 15 J = 1, N2
         ARG = KK*L/N
         CALL ELP(ARG,KC)
         ZR(J) = 0.0
         IF (L.NE.0) ZI(J) = WS/SN
         PR(J) = -WP*SM*CM*CN*DN/(1-((DN*SM)**2.0))
         PI(J) =  WP*DM*SN/(1-((DN*SM)**2.0))
         L = L + 2
 15      CONTINUE
      RETURN
      END
C----------PREWARP OF FREQS BEFORE BLT--------------------
      FUNCTION PREWRP(WW)
      COMMON /PARM/R1,R2,WP,WS,N2,N,SR,KA,KAD,KOD,KF
      IF (KAD.EQ.1) PREWRP = WW
      IF (KAD.NE.1) PREWRP = 2.0*SR*TAN(WW/2.0/SR)
      RETURN
      END




C----------DIGITAL BILINEAR TRANSFORMATION----------------
      SUBROUTINE BLT(N2,SR,R,I)
      REAL SR, R(1), I(1)
C
      A = 2.0*SR
      DO 15 J=1, N2+1
         TR = R(J)
         TI = I(J)
            IF (ABS(TI).GT.1E15) GOTO 10
            IF (ABS(TR).GT.1E15) GOTO 10
         TD = (A - TR)**2 + TI*TI
         R(J) = (A*A - TR*TR - TI*TI)/TD
         I(J) = 2.0*A*TI/TD
         GOTO 15
```

322

```
10          R(J) = -1.0
            I(J) =  0.0
15      CONTINUE
        RETURN
        END
C-------------FREQUENCY TRANSFORMATION--------------------
        SUBROUTINE FREQXFM(W0,PR,PI)
        REAL PR(1), PI(1)
        COMPLEX PC, SC
        COMMON /PARM/R1,R2,WP,WS,N2,N,SR,KA,KAD,KOD,KF
C
        NT = 2*N2+1
        IF (KF.GE.3) GOTO 12
5       DO 10 J=1, N2
            IF (PI(J).GT.1E15) GOTO 7
            PC = CMPLX (PR(J), PI(J))
            SC = 1.0/PC
            PR(J) = -ABS(REAL(SC))
            PI(J) =  ABS(AIMAG(SC))
            GOTO 10
7           PR(J) = 0.0
            PI(J) = 0.0
10      CONTINUE
        RETURN
12      DO 14 J=1, N2
            IF (PI(J).GT.1E15) GOTO 13
            PC = CMPLX (PR(J), PI(J))
            IF (KF.EQ.4) PC = 1.0/PC
            SC = (PC - CSQRT(PC*PC-4*W0*W0))/2.0
            PR(J) = -ABS(REAL(SC))
            PI(J) =  ABS(AIMAG(SC))
            SC = (PC + CSQRT(PC*PC-4*W0*W0))/2.0
            PR(NT-J) = -ABS(REAL(SC))
            PI(NT-J) =  ABS(AIMAG(SC))
            GOTO 14
13          PR(J) = 0.0
            PR(NT-J) = 0.0
            PI(J) = 1E17
            PI(NT-J) = 0.0
            IF (KF.EQ.4) PI(J) = W0
            IF (KF.EQ.4) PI(NT-J) = W0
14      CONTINUE
        RETURN
        END
C-----------COMPLETE ELLIPTIC INTEGRAL---------------
        FUNCTION CEI(KC)
        REAL KC
C
        A = 1.0
        B = KC
        DO 10 J = 1, 20
            AT = (A+B)/2
            B  = SQRT(A*B)
            A  = AT
            IF((A-B)/A .LT. 1.2E-7) GO TO 15
10      CONTINUE
        PRINT *,'CEI FAILED TO CONVERGE'
15      CEI = 1.570796326794896/A
        RETURN
        END
C-------------ELLIPTIC FUNCTIONS------------------
        SUBROUTINE ELP(X,KC)
        REAL KC
        DIMENSION AA(16),BB(16)
        COMMON /ELP1/SN,CN,DN
```

```
C
        IF (X.EQ.0) GOTO 20
        I = 1
        A = 1.0
        B = KC
4       CONTINUE
            AA(I) = A
            BB(I) = B
            AT = (A+B)/2
            B  = SQRT(A*B)
            A  = AT
            IF(((A-B)/A) .LT. 1.3E-7) GO TO 15
            IF (I.GT.15) GOTO 10
            I = I + 1
        GOTO 4
10      PRINT *,'ELP FAILED TO CONVERGE'
15      C = A/TAN(X*A)
        D = 1.0
16      CONTINUE
            E = C*C/A
            C = C*D
            A = AA(I)
            D = (E+BB(I))/(E+A)
            I = I-1
        IF(I.NE.0) GO TO 16
        SN = 1/SQRT(1+C*C)
        CN = SN*C
        DN = D
        RETURN
20          SN = 0.0
            CN = 1.0
            DN = 1.0
        RETURN
        END




C--------------ARC ELLIPTIC TANGENT----------------
        FUNCTION ARCSC(U,KC)
        REAL KC
        A = 1.0
        B = KC
        Y = 1.0/U
        L = 0
        DO 10 J = 1, 15
            BT = A*B
            A  = A + B
            B = 2.0*SQRT(BT)
            Y  = Y - BT/Y
              IF (Y.EQ.0) Y = SQRT(BT)*1E-10
            IF (ABS(A-B).LT.(A*1.2E-7)) GOTO 15
            L = 2*L
            IF (Y.LT.0) L = L + 1
10      CONTINUE
        PRINT *,'ARCSC FAILED TO CONVERGE'
        GOTO 16
15      IF (Y.LT.0) L = L + 1
16      ARCSC = (ATAN(A/Y) + 3.141592654*L)/A
        RETURN
        END
C--------------MODULUS FROM RATIO OF K/K'---------------
        FUNCTION FK(U)
C
        Q = EXP(-3.141592654*U)
        A = 1.0
        B = 1.0
        C = 1.0
        D = Q
        DO 10 J = 1, 15
```

```
                A = A + 2*C*D
                C = C*D*D
                B = B + C
                D = D*Q
                IF (C.LT.1E-7) GOTO 15
     10     CONTINUE
            PRINT *,'FK FAILED TO CONVERGE'
     15     FK = 4*SQRT(Q)*(B/A)**2
            RETURN
            END
C----------------PRINT--------------------------------
            SUBROUTINE PRNT(N2,PR,PI,ZR,ZI)
            DIMENSION PR(20),PI(20),ZR(20),ZI(20)
C
            PRINT *,' #,    ZEROS (REAL, IMAG),          POLES (REAL, IMAG)'
            DO 1 I=1,N2+1
                WRITE (*,10) I,ZR(I),ZI(I),PR(I),PI(I)
     1      CONTINUE
     10     FORMAT (I3,4F14.6)
            RETURN
            END




C-----------CASCADE STRUCTURE PARRAMETERS--------------
            SUBROUTINE CASCAD(PR,PI,ZR,ZI,B1,B2,A1,A2)
            DIMENSION PR(1),PI(1),ZR(1),ZI(1)
            DIMENSION B1(20),B2(20),A1(20),A2(20)
            COMMON /PARM/R1,R2,WP,WS,N2,N,SR,KA,KAD,KOD,KF
C
            PRINT *, N2,' CASCADE STAGES, EACH OF THE FORM:'
            PRINT *,'F(z) = (z*z  +  B1 z  +  B2)/(z*z  +  A1 z  +  A2)'
            K = 0
            IF ((MOD(N2,2).NE.0).AND.(KF.EQ.3)) K = 1
            J0 = 1
            IF (KOD.EQ.0) GOTO 10
                B1(1) =  1.0
                IF (KF.EQ.2) B1(1) = -1.0
                B2(1) =  0.0
                A1(1) = -PR(1)
                A2(1) =  0.0
                WRITE (*,30) J0, B1(J0), B2(J0), A1(J0), A2(J0)
                J0 = 2
     10     DO 15 J = J0, N2
                B1(J) = -2.0*ZR(J)
                B2(J) =  ZR(J)*ZR(J) + ZI(J)*ZI(J)
                    IF ((J.EQ.1).AND.(K.EQ.1)) B1(J) = 0.0
                    IF ((J.EQ.1).AND.(K.EQ.1)) B2(J) = -1.0
                A1(J) = -2.0*PR(J)
                A2(J) =  PR(J)*PR(J) + PI(J)*PI(J)
                    IF (PI(J).EQ.0) A1(1) = -PR(1)-PR(N2+1)
                    IF (PI(J).EQ.0) A2(1) =  PR(1)*PR(N2+1)
                WRITE (*,30) J, B1(J), B2(J), A1(J), A2(J)
     15     CONTINUE
            RETURN
     30     FORMAT ('     ',I3,2F12.6,'   ',2F12.6)
            END
C--------------ANALOG FILTER FREQ RESPONSE--------------------
            SUBROUTINE AFR
            PRINT *,'ANALOG PART NOT FINISHED'
            RETURN
            END
```

```
C------------DIGITAL FILTER FREQ RESPONSE--------------------
      SUBROUTINE DFR(KK,B1,B2,A1,A2,FM)
      DIMENSION B1(1),B2(1),A1(1),A2(1)
      DIMENSION FM(1)
      COMMON /PARM/R1,R2,WP,WS,N2,N,SR,KA,KAD,KOD,KF
C
      Q = 3.141592654/KK
      DO 20 J = 1,  KK+1
         W  = Q*(J-1)
         W2 = 2.0*W
         BR = 1.0
         BI = 0.0
         AR = 1.0
         AI = 0.0
         I0 = 1
         IF (KOD.EQ.0) GOTO 10,
            BR = COS(W)  + B1(1)
            BI = SIN(W)
            AR = COS(W)  + A1(1)
            AI = SIN(W)
            I0 = 2
10       DO 15 I = I0, N2
            BRT = COS(W2) + B1(I)*COS(W)  + B2(I)
            BIT = SIN(W2) + B1(I)*SIN(W)
            ART = COS(W2) + A1(I)*COS(W)  + A2(I)
            AIT = SIN(W2) + A1(I)*SIN(W)
            BRS = BR*BRT - BI*BIT
            BI  = BR*BIT + BI*BRT
            BR  = BRS
            ARS = AR*ART - AI*AIT
            AI  = AR*AIT + AI*ART
            AR  = ARS
15       CONTINUE
         FM(J) = SQRT((BR*BR + BI*BI)/(AR*AR + AI*AI))
20    CONTINUE
      RETURN
      END
C-----------OUTPUT FREQUENCY RESPONSE------------------
      SUBROUTINE VIEW(KK,FM)
      DIMENSION FM(1)
C
      OPEN (1,FILE='fm')
      REWIND(1)
      F0 = 0.5/KK
      DO 10 J = 1, KK+1
         F = F0*(J-1)
         WRITE (1,100) F, FM(J)
10    CONTINUE
100   FORMAT(10X,F15.8,E15.8)
      RETURN
      END
```

REFERENCES

[21] M. Abramowitz and I. A. Stegun, eds., *Handbook of Mathematical Functions*, Washington, D.C.: National Bureau of Standards, 1964, Chaps. 16 and 17.

[22] R. Bulirsch, "Numerical Calculation of Elliptic Integrals and Elliptic Functions," *Numer. Math.* **7** 78–90 (1965).

[23] A. H. Gray and J. D. Markel, *IEEE Trans ASSP*, 1976.

326

10. A FORTRAN PROGRAM FOR LOW-PASS FIR FILTER DESIGN USING A LEAST SQUARED EQUATION-ERROR CRITERION

The FORTRAN program for designing digital IIR filters uses a discrete LS equation-error criterion. In the limit, where the number of specified frequency-response points is equal to the number of filter coefficients, the method becomes a frequency-sampling design technique. The basic theory, formulas, and variable names are chosen to follow the development in Section 7.4.2.

The main program starts with a section that takes input specifications from the terminal. A subroutine calculates the frequency response of the finished filter design at a specified number of equally spaced points for display on the terminal and writes them into a file for further use. The number of frequency points to be evaluated is entered first. The transfer function is assumed to be of the form

$$F(z) = \frac{b_0 z^M + b_1 z^{M-1} + b_2 z^{M-2} + \cdots + b_M}{(z^N + a_1 z^{N-1} + \cdots + a_N)z^{M-N}}$$

as in equation (7.116). The order of the numerator M and the order of the denominator N are entered next. Note that the number of unknown coefficients in the numerator is $M + 1$ and in the denominator is N. Then the number of frequency samples to be approximated is entered as L1 and the band edge as FP in Hertz. The minimum number of frequencies that will uniquely determine the $M + N + 1$ filter coefficients is $M + N + 1$. See Section 7.4.1.

The next section sets the L1 desired frequency-response samples in two arrays, with C(J) being the samples of the real part and D(J) being the samples of the imaginary part. Note that the FORTRAN indices start at 1 rather than 0 as the equations do. The program assumes real filter coefficients. Therefore, the real part of the frequency response is even, and the imaginary part is odd. This fact, coupled with the fact that the frequency response of all digital filters is periodic, means that C(J) = C(L1 + 1 − J) and D(J) = −D(L1 + 1 − J). This also implies that, for all cases, D(1) = 0, and, for even L1, D(L1/2 + 1) = 0. The number of frequency samples is approximately L1/2, but because the samples are complex valued, the number of real values required is L1. For example, if L1 = 5, there will be three independent real-part samples to specify and two independent imaginary-part samples. For L1 = 4 there will be three independent real-part samples and one imaginary-part sample.

The next section of the program takes the IDFT of the frequency-sample vector as required in (7.120) and (7.121). The subroutine IDFT( ) takes into account that the coefficients are real and the frequencies have symmetries.

The next two sections, in the DO 20 and DO 21 loops, form the H_1, H_2, and h_1 matrices described in (7.124). They are labeled H1, H2, and H0 in the program. Equation (7.125) or (7.128) is solved by the subroutines from the software package LINPACK.[9] If L1 = $M + N + 1$, it is a matrix inversion. If L1 > $M + N + 1$, a LS equation error is found, as in (7.130), by LINPACK.

Note that this does not give a LS error approximation to the desired frequency response, but a LS error solution of equation (7.127).

In the next two sections the normalized A0 coefficient is appended to the A(J) vector, and the B(J) vector is calculated from (7.126). The actual design process is now complete. An analysis section follows, which calculates the frequency response from the filter coefficients in A and B by using subroutine DFR( ). The resulting K frequency-response values are written to the file fm by subroutine VIEW( ).

This program is a straightforward implementation of the theory in Sections 7.4.1 and 7.4.2. The input and output are primitive and need to be customized by the user. The various modifications described in 7.4.2 can easily be added. Recall that this procedure does not minimize the usual solution error described in (7.132) or (7.133). It minimizes the squared equation error defined in (7.128). One of the powerful features of this approach is the approximation of a complex desired frequency response. Experience shows that very surprising results are often obtained if the desired frequency response is not close to what an IIR filter of the specified order can achieve. The inclusion of phase specifications is a significant complication compared to magnitude-only approximations. A root-finder subroutine could be added to factor the numerator and denominator of $H(z)$, and the cascade structure section from Program 8 could easily be added.

A program to implement Prony's method for time-domain design of IIR filters, as described in Section 7.5, could easily be written by modifying this program. The formation of the basic matrices in (7.138) and their solution in (7.139) and (7.140) are the same as done here. The input section would have to be changed and the IDFT would have to be removed.

```
C----------------------------------------------------------------
C        A FREQUENCY-SAMPLING AND DISCRETE
C        LEAST-SQUARED-EQUATION-ERROR IIR FILTER DESIGN
C        PROGRAM. REQUIRES LINPACK.
C          NUMERATOR ORDER: M;      M+1 COEFFS:  B(K)
C          DENOMINATOR ORDER: N;    N+1 COEFFS:  A(K), A(1) =1
C          NUMBER OF FREQ SAMPLES: L+1
C          FREQ SAMPLE METHOD:   L = M + N
C          DISCRETE LEAST SQR:   L > M + N
C        C.S. BURRUS,  RICE UNIV.,  FEB 1986
C----------------------------------------------------------------
         REAL A(50),B(50),C(501),D(501),H(501),H0(501),QAX(50)
         REAL FM(530)
         REAL H1(50,50), H2(501,50)
C
         LDX = 501
         PRINT *,'FS AND LS  DESIGN OF AN IIR FILTER'
         PRINT *,'ENTER NUMBER OF FREQUENCIES TO BE EVALUATED'
         READ *, KK
  1      PRINT *,'ENTER NUMERATOR ORDER, DENOMINATOR ORDER'
         READ *, M, N
         PRINT *,'ENTER THE NUMBER OF FREQ SAMPLES TO OPTIMIZE OVER'
         READ *, L1
```

```
C
       L  = L1- 1
       LM = L - M
       M1 = M + 1
       N1 = N + 1
       ML = (L1+1)/2
C---------------SET THE DESIRED FREQUENCY RESPONSE----
       PRINT *,'ENTER THE ',ML,' REAL PART SAMPLES'
       READ *, (C(J),J=1,ML)
       PRINT *,'ENTER THE ',ML,' IMAG PART SAMPLES'
       READ *, (D(J),J=1,ML)
C---------------TAKE THE INVERSE DFT------------------
       CALL IDFT(L1,C,D,H)
C---------------FORM THE H1 MATRIX--------------------
       DO 20 J = 1, M1
          I = J
          DO 10 K = 1, N1
             IF (I.LT.1)  I = L1
             H1(J,K) = H(I)
             I = I - 1
  10      CONTINUE
  20   CONTINUE
C-----------------FORM THE H2 AND H0 MATRICES----------
       I0 = M1
       DO 21 J = 1, LM
          H0(J) = -H(I0+1)
          I = I0
          DO 11 K = 1, N
             IF (I.LT.1) I = L1
             H2(J,K) = H(I)
             I = I - 1
  11      CONTINUE
          I0 = I0 + 1
  21   CONTINUE




C---------------LEAST SQUARES SOLUTION FROM LINPAC----
       CALL SQRDC(H2,LDX,LM,N,QAX,DUM,DUM,0)
       CALL SQRSL(H2,LDX,LM,N,QAX,H0,DUM,H0,A,DUM,DUM,100,INFO)
C---------------ADD THE UNITY TERM TO A---------------
       DO 25 J = 1, N
          A(N+2-J) = A(N+1-J)
  25   CONTINUE
       A(1) = 1.0
C---------------CALCULATE THE NUMERATOR COEFFS--------
       DO 40 J = 1, M1
          BT = 0.0
          DO 30 K = 1, N1
             BT = BT + H1(J,K)*A(K)
  30      CONTINUE
          B(J) = BT
  40   CONTINUE
C-------------OUTPUT COEFFS AND FREQ RESPONSE---------
       PRINT *,'NUMERATOR COEFFS ARE:'
       PRINT *, (B(J),J=1,M1)
       PRINT *,'DENOMINATOR COEFFS ARE:'
       PRINT *, (A(J),J=1,N1)
       CALL DFR(N,M,KK,A,B,FM)
       CALL VIEW(KK,FM)
       GOTO 1
       END
```

```
C-------------END OF MAIN PROGRAM--------------------
C
        SUBROUTINE DFR(N,M,KK,A,B,FM)
        REAL A(1), B(1), FM(1)
        Q = 3.141592654/KK
        DO 20 J = 1, KK+1
            BR = B(M+1)
            BI = 0.0
            QQ = Q*(J-1)
            DO 10 I = 1, M
                BR = BR + B(M+1-I)*COS(QQ*I)
                BI = BI + B(M+1-I)*SIN(QQ*I)
    10      CONTINUE
            AR = A(N+1)
            AI = 0.0
            DO 15 I = 1, N
                AR = AR + A(N+1-I)*COS(QQ*I)
                AI = AI + A(N+1-I)*SIN(QQ*I)
    15      CONTINUE
            FM(J) = SQRT((BR*BR + BI*BI)/(AR*AR + AI*AI))
    20  CONTINUE
        RETURN
        END




C-------------------------------------------------------
        SUBROUTINE IDFT(N,C,D,H)
        REAL C(1), D(1), H(1)
C
        Q  = 6.283185307179586/N
        M  = (N+1)/2
        DO 20 J = 1, N
            HT = 0.0
            DO 10 K = 2, M
                QQ = Q*(J-1)*(K-1)
                HT = HT + C(K)*COS(QQ) - D(K)*SIN(QQ)
    10      CONTINUE
            HT = C(1) + 2*HT
            IF (MOD(N,2).EQ.0) HT = HT + C(M+1)*COS(3.141592654*(J-1))
            H(J) = HT/N
    20  CONTINUE
        RETURN
        END
C----------------------------------------------------------
        SUBROUTINE VIEW(KK,FM)
        DIMENSION FM(1)
        OPEN (1,FILE='fm')
        REWIND (1)
        DO 10 J = 1, KK
            F = 0.5*(J-1)/KK
            WRITE (1,100) F,FM(J)
    10  CONTINUE
   100  FORMAT (10X,F15.8,E15.8)
        RETURN
        END
```

11–13. TMS32010 ASSEMBLY LANGUAGE PROGRAMS

This section contains three assembly language programs for the TMS32010. Program FIR21 implements the length-21 FIR filter described in the design example in Chapter 5. It is written to be run on the Texas Instruments (TI) EVM evaluation module board in conjunction with the TI AIB analog interface board. For details on calculation and scaling of the coefficients, see Chapter 5.

The second and third programs are implementations of the fourth-order elliptic filter described in the design example in Chapter 8. These programs are written to be run on the simulator provided by TI for the TI or IBM PC. A few additional instructions are required to run these programs on the EVM/AIB boards.

The second program uses the transpose structure for each of the cascaded second-order blocks, and the third program uses the direct structure for each of the cascaded second-order blocks. We were careful to avoid overflow limit cycles, by using the overflow mode of arithmetic. This required that the output be scaled up at the end of the programs with the APAC instruction, because of the implementation of overflow detection on the TMS32010. If the same filter were implemented on the TMS32020, these extra instructions would not be necessary. For details on calculation and scaling of the coefficients, see Chapter 8.

```
            IDT       'FIR21'
*
*************************
*  NAME DATA LOCATIONS  *
*************************
*
CLCK    EQU      0
MODE    EQU      1
MASK    EQU      2
SIGN    EQU      3
CONV    EQU      4
ONE     EQU      5
YN      EQU      6
X1      EQU      11
X2      EQU      12
X3      EQU      13
X4      EQU      14
X5      EQU      15
X6      EQU      16
X7      EQU      17
X8      EQU      18
X9      EQU      19
X10     EQU      20
X11     EQU      21
X12     EQU      22
X13     EQU      23
X14     EQU      24
X15     EQU      25
X16     EQU      26
X17     EQU      27
X18     EQU      28
X19     EQU      29
X20     EQU      30
X21     EQU      31
H1      EQU      51
H2      EQU      52
H3      EQU      53
H4      EQU      54
H5      EQU      55
H6      EQU      56
H7      EQU      57
H8      EQU      58
H9      EQU      59
H10     EQU      60
H11     EQU      61
*
```

```
************************
*  START OF PROGRAM   *
************************
*
        AORG    0
        B       START           BRANCH AROUND DATA
*
***     SET DATA    ***
*
DCLCK   DATA    >200            PROGRAM PARAMETERS
DMODE   DATA    >8
DMASK   DATA    >FFF0
DSIGN   DATA    >8000
DCONV   DATA    >4000
DONE    DATA    >1
DH1     DATA    >011D           FILTER COEFFICIENT
        DATA    >035D               "
        DATA    >FD82               "
        DATA    >00E9               "
        DATA    >01AE               "


        DATA    >FC62               "
        DATA    >02BC               "
        DATA    >01F2               "
        DATA    >F60F               "
        DATA    >0FAA               "
        DATA    >2AAF               "
*
**************************
*  INITIALIZATION CODE   *
**************************
*
START   LDPK    0               SET DATA PAGE POINTER
        SOVM
        DINT
        LACK    DCLCK           READ IN PROGRAM PARAMETERS
        TBLR    CLCK                "
        LACK    DMODE
        TBLR    MODE
        LACK    DMASK
        TBLR    MASK                "
        LACK    DSIGN
        TBLR    SIGN
        LACK    DCONV
        TBLR    CONV                "
        LACK    DONE
        TBLR    ONE
*
        LACK    DH1             READ IN FILTER COEFFICIENTS
        TBLR    H1                  "
        ADD     ONE
        TBLR    H2
        ADD     ONE
        TBLR    H3                  "
        ADD     ONE
        TBLR    H4                  "
        ADD     ONE
        TBLR    H5
        ADD     ONE
        TBLR    H6                  "
        ADD     ONE
        TBLR    H7                  "
        ADD     ONE
        TBLR    H8
        ADD     ONE                 "
        TBLR    H9
        ADD     ONE
        TBLR    H10                 "
        ADD     ONE
        TBLR    H11
*
```

```
********************
*  INITIALIZE AIB  *
********************
*
        OUT     CLCK,PA1            SET SAMPLING RATE
        OUT     MODE,PA0            SET AIB MODE
*
***************************
*  WAIT FOR NEXT SAMPLE   *
***************************
*
FILT    BIOZ    GET                BRANCH ON NEW SAMPLE
        B       FILT
*
***************************
*  IMPLEMENT THE FILTER   *


***************************
*
GET     IN      X1,PA2             READ IN SAMPLE
        LAC     X1                 CONVERT SAMPLE TO TWO'S COMPLEMENT
        XOR     MASK                            "
        AND     MASK
        ADDS    SIGN
        SACL    X1
*
        ZAC                        COMPUTE NEXT OUTPUT
        LT      X21
        MPY     H1                          "
        LTD     X20
        MPY     H2
        LTD     X19
        MPY     H3
        LTD     X18
        MPY     H4
        LTD     X17
        MPY     H5                          "
        LTD     X16
        MPY     H6
        LTD     X15
        MPY     H7
        LTD     X14
        MPY     H8
        LTD     X13                         "
        MPY     H9
        LTD     X12
        MPY     H10
        LTD     X11                         "
        MPY     H11
        LTD     X10
        MPY     H10
        LTD     X9
        MPY     H9
        LTD     X8                          "
        MPY     H8
        LTD     X7
        MPY     H7
        LTD     X6                          "
        MPY     H6
        LTD     X5
        MPY     H5
        LTD     X4                          "
        MPY     H4
        LTD     X3
        MPY     H3
        LTD     X2                          "
        MPY     H2
        LTD     X1
        MPY     H1
        APAC
```

```
        ADD     CONV,15         CONVERT TO BINARY FORMAT
        SACH    YN,1
        OUT     YN,PA2  OUTPUT Y(N)
        B       FILT            WAIT FOR NEXT SAMPLE
        END  ·
```


```
***************************************************
                    PROGRAM 12
***************************************************


***************************************************
*                                                 *
*  FOURTH ORDER ELLIPTIC LOWPASS FILTER, TWO CASCADED *
*  BIQUAD SECTIONS (TRANSPOSE STURCTURE).  FILTER *
*  COEFFICIENTS OF EACH SECOND-ORDER SECTION ARE  *
*  SCALED BY THE LARGEST L1 NORM                  *
*  OF THE IMPULSE RESPONSE OF EACH SUMMING NODE, THUS *
*  GUARANTEEING NO OVERFLOW.  THE FILTER OUTPUT IS *
*  SCALED UP TO IMPLEMENT AN OVERALL GAIN OF ONE, AND *
*  THUS THERE IS A POSSIBILITY OF OVERFLOW THERE. THE *
*  UNSCALED REPRESENTATION OF THE OUTPUT (BEFORE THE *
*  MULTIPLICATION BY H0) MAY BE USED TO GUARANTEE NO *
*  OVERFLOW AT THE FILTER OUTPUT AS WELL.         *
*                                                 *
*  THE NOTATION (Q15*2) INDICATES THAT THE COEFFICIENT *
*  IN THE PROGRAM IS TWICE THE HEX EQUIVALENT OF THE *
*  DECIMAL NUMBER.                                *
*  IN THIS PROGRAM, THE SECOND INDEX OF COEFFICIENTS *
*  AND VARIABLES INDICATES THE SECTION THE COEFFICIENT *
*  (VARIABLE) BELONGS TO,EG. A21 IS THE COEFFICIENT OF *
*  Z**-2 FOR THE FIRST STAGE. IN CHAPTER 8, THE   *
*  OPPOSITE CONVENTION IS USED,I.E., a12 IS       *
*  THE COEFFICIENT OF Z**-2 IN THE TEXT.          *
***************************************************
*
XN      EQU     0
YN1     EQU     1
Y21     EQU     2
Y11     EQU     3
YN2     EQU     4
Y22     EQU     5
Y12     EQU     6
A11     EQU     7
A21     EQU     8
B01     EQU     9
B11     EQU     10
B21     EQU     11
A12     EQU     12
A22     EQU     13
B02     EQU     14
B12     EQU     15
B22     EQU     16
H0      EQU     17
ONE     EQU     18
*
        AORG    0
RSLOC   B       INIT
*
```

334

```
            DATA    >6730       -A11  =    0.4030703       (Q15  *  2)
            DATA    >C449       -A21  =   -0.2332662       (Q15  *  2)
            DATA    >303C        B01  =    0.1884133       (Q15  *  2)
            DATA    >4E3A        B11  =    0.3055656       (Q15  *  2)
            DATA    >303C        B21  =    0.1884133       (Q15  *  2)
*




            DATA    >F2D6       -A12  =   -0.0514214       (Q15  *  2)
            DATA    >99F3       -A22  =   -0.7972861       (Q15)
            DATA    >59A9        B02  =    0.3502291       (Q15  *  2)
            DATA    >4030        B12  =    0.2507274       (Q15  *  2)
            DATA    >59A9        B22  =    0.3502291       (Q15  *  2)



*
            DATA    >476F        H0
*
WONE        DATA    >0001
*
INIT        LDPK    0
            SOVM
            LARK    AR0,ONE
            LARK    AR1,11
            LACK    WONE
TABLER      LARP    AR0
            TBLR    *-,AR1
            SUB     ONE
            BANZ    TABLER
*
ZFILT       ZAC
            SACL    Y21
            SACL    Y11
            SACL    Y22
            SACL    Y12
*
IIR2TS      IN      XN,PA0
LPSEC1      LT      XN          Q15
            MPY     B01         Q15*2
            ZALH    Y21         Q15
            APAC
            SACH    YN1         Q15
            MPY     B11         Q15*2
            ZALH    Y11         Q15
            LTA     YN1         Q15
            MPY     A11         Q15*2
            APAC
            SACH    Y21         Q15
            MPY     A21         Q15*2
            PAC
            LT      XN          Q15
            MPY     B21         Q15*2
            APAC
            SACH    Y11         Q15
*
LPSEC2      LT      YN1         Q15
            MPY     B02         Q15*2
```

```
      ZALH  Y22      Q15
      APAC
      SACH  YN2      Q15
      MPY   B12      Q15*2
      ZALH  Y12      Q15
      LTA   YN2      Q15
      MPY   A12      Q15*2
      APAC
      SACH  Y22      Q15
      MPY   A22      Q15
      PAC
      APAC           MAKE A22 Q15*2
      LT    YN1      Q15
      MPY   B22      Q15*2
      APAC
      SACH  Y12      Q15
*
      LT    YN2      Q15
      MPY   H0       Q15/4
      PAC
      APAC
      APAC           SCALE YN BACK UP -- NOTE THAT
      APAC           OVERFLOW (SATURATION) MAY OCCUR
      APAC           HERE IF OUTPUT MAGNITUDE EXCEEDS 1
      APAC
      APAC
      APAC
      SACH  YN2      Q15
*
      OUT   YN2,PA1
      B     IIR2TS
*
      END
*
*************************************************************
                      PROGRAM 13
*************************************************************

*************************************************************
*                                                          *
* FOURTH ORDER ELLIPTIC LOWPASS FILTER, TWO CASCADED       *
* BIQUAD (DIRECT FORM 2 STRUCTURE) SECTIONS.  FILTER       *
* COEFFICIENTS OF EACH SECOND-ORDER SECTION SCALED BY      *
* THE LARGEST L1 NORM OF THE IMPULSE RESPONSE OF EACH      *
* SUMMING NODE, THUS SUMMING NODE, THUS GUARANTEEING       *
* NO OVERFLOW.  THE FILTER OUTPUT IS SCALED UP TO          *
* IMPLEMENT AN OVERALL GAIN OF ONE, AND THUS THERE IS      *
* A POSSIBILITY OF OVERFLOW THERE. THE UNSCALED            *
* REPRESENTATION OF THE OUTPUT (BEFORE THE                 *
* MULTIPLICATION BY H0) MAY BE USED TO GUARANTEE NO        *
* OVERFLOW AT THE FILTER OUTPUT AS WELL.                   *
*                                                          *
* THE NOTATION (Q15*2) INDICATES THAT THE COEFFICIENT      *
* IN THE PROGRAM IS TWICE THE HEX EQUIVALENT OF THE        *
* DECIMAL NUMBER.                                          *
*                                                        *
*                                                        *
* IN THIS PROGRAM, THE SECOND INDEX OF COEFFICIENTS        *
* AND VARIABLES INDICATES THE SECTION THE COEFFICIENT      *
* (VARIABLE) BELONGS TO,EG. A21 IS THE COEFFICIENT OF      *
```

```
*   Z**-2 FOR THE FIRST STAGE. IN CHAPTER 8, THE        *
*   OPPOSITE CONVENTION IS USED,I.E., a12 IS           *
*   THE COEFFICIENT OF Z**-2 IN THE TEXT.              *
**********************************************************
*
XN        EQU    0
YN1       EQU    1
Y01       EQU    2
Y11       EQU    3
Y21       EQU    4
YN2       EQU    5
Y02       EQU    6
Y12       EQU    7
Y22       EQU    8
SCLFAC    EQU    9
A11       EQU    10
A21       EQU    11
B01       EQU    12
B11       EQU    13
B21       EQU    14
A12       EQU    15
A22       EQU    16
B02       EQU    17
B12       EQU    18
B22       EQU    19
H0        EQU    20
ONE       EQU    21
*
          AORG   0
RSLOC     B      INIT
*
          DATA   >5D19      SCLFAC =  0.3636636     (Q15 * 2)
*
          DATA   >6730      -A11  =   0.4030703     (Q15 * 2)
          DATA   >C449      -A21  =  -0.2332662     (Q15 * 2)
          DATA   >1411      B01   =   0.0783843     (Q15 * 2)
          DATA   >208B      B11   =   0.1271224     (Q15 * 2)
          DATA   >1411      B21   =   0.0783843     (Q15 * 2)
*
          DATA   >F2D6      -A12  =  -0.0514214     (Q15 * 2)
          DATA   >99F3      -A22  =  -0.7972861     (Q15)
          DATA   >7FFF      B02   =   0.9999695     (Q15)
          DATA   >5BA2      B12   =   0.7158737     (Q15)
          DATA   >7FFF      B22   =   0.9999695     (Q15)
*
          DATA   >52AF      H0    =   5.1676996     (Q15/8)
*
WONE      DATA   >0001
*
INIT      LDPK   0
          SOVM
          LARK   AR0,ONE
          LARK   AR1,12
          LACK   WONE
TABLER    LARP   AR0
          TBLR   *-,AR1
          SUB    ONE
          BANZ   TABLER
```

337

```
*
ZFILT   ZAC
        SACL    Y01
        SACL    Y11
        SACL    Y21
        SACL    Y02
        SACL    Y12
        SACL    Y22
*
IIR2DS  IN      XN,PA0
        LT      XN
        MPY     SCLFAC      Q15*2
LPSEC1  PAC
        LT      Y11         Q15
        MPY     A11         Q15*2
        LTA     Y21         Q15
        MPY     A21         Q15*2
        APAC
        SACH    Y01         Q15
        ZAC
        MPY     B21         Q15*2
        LTD     Y11
        MPY     B11         Q15*2
        LTD     Y01
        MPY     B01         Q15*2
*
LPSEC2  LTA     Y12         Q15
        MPY     A12         Q15*2
        LTA     Y22         Q15
        MPY     A22         Q15
        APAC
        APAC                MAKE A22 Q15*2
        SACH    Y02         Q15
        MPY     B22         Q15
        PAC
        LTD     Y12         MAKE B22 Q15*2
        MPY     B12
        APAC
        LTD     Y02         MAKE B12 Q15*2
        MPY     B02
        APAC
        APAC                MAKE B02 Q15*2
        SACH    YN2
*
        LT      YN2         Q15
        MPY     H0          Q15/8
        PAC
        APAC
        APAC                SCALE YN BACK UP -- NOTE THAT
        APAC                OVERFLOW (SATURATION) MAY OCCUR
        APAC                HERE IF OUTPUT MAGNITUDE EXCEEDS 1
        APAC
        APAC
        APAC
        APAC
        APAC
        APAC
        APAC
        APAC
```

```
          APAC
          APAC
          APAC
          SACH    YN2         Q15
*
          OUT     YN2,PA1
          B       IIR2DS
*
          END
```

Index

QUEEN MARY
COLLEGE
LIBRARY

VISTECH

2 ARLINGTON DRIVE

WOODSMOOR

STOCKPORT

CHESHIRE   SK2 7PB

Tel 061 - 483   5176

IEE SP 406 1992
Buses

ω

ω S

$$S + \frac{\omega_0}{S}$$

$$\frac{\omega_0^2 - \omega^2}{i\omega}$$

$$-i\frac{\omega_0^2 - \omega}{\omega}$$

$$\frac{\omega^2 - \omega_0^2}{\omega} = 0$$

$$= \frac{\omega^2 - \omega_0^2}{\omega} c$$

$$i\omega - i\frac{\omega_0^2}{\omega}$$

$$= i\left(\omega - \frac{\omega_0^2}{\omega}\right)$$

$$\omega = \frac{\omega_0^2}{\omega}$$

$$\omega^2 = \omega_0^2$$